Waste and Waste Management

nova
science publishers
www.novapublishers.com

Waste and Waste Management

Development of End-Markets for Recycled Construction and Demolition Waste Resources in Australia
Salman Shooshtarian, Tayyab Maqsood,
Savindi Caldera, Tim Ryley and Peter S. P. Wong
2023. ISBN: 979-8-88697-956-5 (Hardcover)
2023. ISBN: 979-8-89113-121-7 (eBook)

Fuel Briquettes Made of Carbon-Containing Technogenic Raw Materials
Nina Buravchuk and Olga Guryanova
2023. ISBN: 979-8-88697-907-7 (Softcover)
2023. ISBN: 979-8-88697-944-2 (eBook)

Water Supply and Drainage for Buildings
Syed Azizul Haq
2023. ISBN: 979-8-88697-607-6 (Hardcover)
2023. ISBN: 979-8-88697-761-5 (eBook)

Municipal Solid Waste Management and Improvement Strategies
Adam Fitz (Editor)
2023. ISBN: 979-8-88697-720-2 (Softcover)
2023. ISBN: 979-8-88697-756-1 (eBook)

Algal Biorefining: Resource Expenditure and Exergo-Environmental Sustainability
Cynthia Ofori-Boateng
2022. ISBN: 978-1-68507-706-8 (Hardcover)
2022. ISBN: 978-1-68507-921-5 (eBook)

More information about this series can be found at
https://novapublishers.com/product-category/series/waste-and-waste-management/

Salman Shooshtarian, PhD
Tayyab Maqsood, PhD
Savindi Caldera, PhD
Tim Ryley, PhD
and Peter S. P. Wong, PhD

Development of End-Markets for Recycled Construction and Demolition Waste Resources in Australia

nova
science publishers
www.novapublishers.com

DOI: https://doi.org/10.52305/PZRQ5282

NOTICE TO THE READER

Library of Congress Cataloging-in-Publication Data

Names: Shooshtarian, Salman, author.
Title: Development of end-markets for recycled construction and demolition waste resources in Australia / Salman Shooshtarian, Tayyab Maqsood, Savindi Caldera, Tim Ryley and Peter SP Wong.
Description: New York : Nova Science Publishers, [2023] | Series: Waste and waste management | Includes bibliographical references and index. |
Identifiers: LCCN 2023035518 (print) | LCCN 2023035519 (ebook) | ISBN 9798886979565 (hardcover) | ISBN 9798891131217 (adobe pdf)
Subjects: LCSH: Construction and demolition debris--Recycling--Australia. | Construction and demolition debris--Australia--Management. | Construction industry--Waste disposal--Australia.
Classification: LCC TD899.C5885 M367 2023 (print) | LCC TD899.C5885 (ebook) | DDC 690.028/60994--dc23/eng/20230811
LC record available at https://lccn.loc.gov/2023035518
LC ebook record available at https://lccn.loc.gov/2023035519

Published by Nova Science Publishers, Inc. † New York

Contents

List of Figures

List of Tables

Preface

A growing population and urbanisation have led to a significant increase in construction and demolition (C&D) waste generation. There are urgent calls for government and industry to develop better methods for C&D waste management. The creation and stimulation of a market for C&D waste PwRC have emerged as targeted interventions to divert waste from landfill sites and create a second life for waste material. In response to industry and public concerns related to unsustainable C&D waste management, this book aims to enhance understanding of the operation and capacity of end-markets for C&D waste recyclables in Australian jurisdictions. This understanding involves the integration of circular economy principles and a cradle-to-cradle approach into C&D waste management systems in the Australian context.

This publication is timely as it focuses on addressing the challenges of poor resource efficiency and inadequate performance in carbon emission mitigation in the architecture, engineering and construction (AEC) industry from a market development perspective. The resource efficiency is recognised as one of the six national research priorities. The book underscores the importance of incorporating circular economy principles in a comprehensive and cross-cutting manner to guide the management of C&D waste.

This book presents an extensive review of various national and international literature, analysis of relevant policies and guidelines, and evaluation of key stakeholders' sentiments. It provides valuable insights into five key areas: (1) analysis of national and jurisdictional policy frameworks that affect creation and stimulation of end-markets for waste materials; (2) assessment of the operation of existing end-markets for five strategic construction materials; (3) exploration of the application of sustainable procurement policies and practices at state, national and international levels; (4) evaluation of impacts of COVID-19 on the C&D waste industry; and (5) identification of key factors affecting C&D waste end-market creation and stimulation, and building a model for waste market development.

The authors of this book are highly experienced academics with a well-established track record of conducting national research in the field of C&D waste management and the circular economy. Over the past six years, the authors have collaborated on three national research projects that have delved into various aspects of sustainable waste management within the Australian AEC industry. The findings and outcomes of this research have been widely disseminated through diverse communication channels, targeting industry, government, academic and public audiences. Additionally, the authors have actively engaged in policy advocacy efforts, making multiple submissions to local, jurisdictional, and national waste policies and strategies calls, further showcasing their commitment to driving positive change in the field of waste management.

Acknowledgments

This research has been developed with support provided by Australia's Sustainable Built Environment National Research Centre (SBEnrc). SBEnrc develops projects informed by industry partner needs, secures national funding, project manages collaborative research and oversees research into practice initiatives. Core Members of SBEnrc include ATCO Australia, BGC Australia, Government of Western Australia, Queensland Government, Curtin University, Griffith University, RMIT University and Western Sydney University. This research would not have been possible without the valuable support of our core industry, government and research partners.

We express our heartfelt gratitude to Gayle Sloan, the CEO of the Australian Waste Management and Resource Recovery Association, for her unwavering support in guiding the SBEnrc's Project (No. 1.85) Steering Group and providing invaluable contacts in the waste management industry for consultation and data collection. Her expertise has greatly contributed to aligning the research direction with the industry's needs and concerns, ensuring a more effective and relevant outcome.

The authors would like to express their appreciation to Associate Professor Atiq Zaman from Curtin University for graciously facilitating interviews with experts in Western Australia. Finally, the authors extend their sincere appreciation to the numerous individuals and organisations who have supported their research and the resulting book. These contributions have been invaluable and have greatly contributed to the successful completion of this project. Thank you for your support and commitment to advancing knowledge in this field.

List of Abbreviations

ABS	Australian Bureau of Statistics
ACT	Australian Capital Territory
AEC	Architecture, Engineering and Construction
Billion	bn
C&D	Construction and Demolition
C&I	Commercial and Industrial
CE	Circular Economy
EfW	Energy from Waste
EoW	End of Waste
EPA	Environmental Protection Authority
GDP	Gross Domestic Product
GHG	Greenhouse Gas Emissions
GISA	Green Industries SA
GPP	Green Public Procurement
GSP	Gross State Product
IPWEA	Institute of Public Works Engineering Australasia
ISO	International Organization for Standardization
kt	Kilo tonne
m	Million
MRF	Material Recycling Facility
MSW	Municipal Solid Waste
Mt	Million tonnes
NSW	New South Wales
NT	Northern Territory
NWP	National Waste Policy
OECD	Organisation for Economic Co-operation and Development
PwRC	Product with Recycled Content
Qld	Queensland
RAP	Reclaimed Asphalt Pavement

SA	South Australia
SP	Sustainable Procurement
SPP	Sustainable Public Procurement
SV	Sustainability Victoria
t	Tonne (metric)
Tas	Tasmania
TfN	Transport for NSW
UN	United Nations
UNEP	United Nations Environment Program
Vic	Victoria
WA	Western Australia
WMD	Waste Management Document

Chapter 1

Introduction

1.1. Introduction

This chapter introduces the structure of this book, describes its background and highlights the observed knowledge gap that the studies included sought to address. The knowledge gap identified directly informed the research objectives, and subsequently the approach, analysis and synthesis of information.

Rapid population growth and urbanisation have led to a significant increase in Construction and Demolition (C&D) waste generation globally [1]. Urbanisation involves creating or maintaining built environments to meet the urban residents' growing needs to live, work and commute. In Australia, C&D waste streams are the largest source of solid waste. The latest data shows that, in the fiscal year 2018–19, about 27 Mt of C&D waste was generated, which was the most significant contributor to the total solid waste (44%) in Australia [2]. This increase could be attributed to the extensive urbanisation in many areas within Australia. As a result, annually, a substantial amount of C&D waste is sent to landfill, causing material values to be lost, excessive extraction of virgin materials used in the architecture, engineering and construction (AEC) industry, and negative social, economic and environmental impacts. In reality, and as indicated in the literature, such growth in C&D waste emerges from low resource efficiency in the AEC industry. It also represents the inadequate waste management of C&D projects.

According to the National Waste Report (2020), in 2018–19, about 67% of the total C&D waste generated by the AEC industry was recovered, which fails to meet the target rates set by individual Australian states and territories (such as 90% in South Australia and 75% in Western Australia) [3]. Furthermore, said rates are significantly lower than in some European countries (Eurostat, 2021) including the Netherlands (100%), Ireland (100%), the UK (99%), Italy (98%) and Denmark (97%). Hence, several scholars [4, 5] and policymakers [6, 7] advocate a circular economy (CE) approach to ensure the value of materials is kept in the economy as long as possible. This concept goes beyond the traditional take-make-dispose approach, where there is a planned end (i.e., waste disposal) to resources used in various industries, which inhibits their implementation of the principles of a CE due to losing the resources.

Across the Australian industry supply chain, the drivers for a CE have gained ground with the increasing consciousness of the fundamental importance of the three pillars of sustainability [5, 8]. Practitioners and researchers agree on the benefits of adopting a CE within the AEC context and widely recognise its potential for shifting the AEC industry towards sustainable practices [9]. The universal potential of jobs emerging from the CE's continued growth and new markets reiterated in national and jurisdictional policies has been substantiated by reports estimating massive gross domestic product (GDP) and consumption boosts due to such an approach [10].

In the C&D waste management context, a critical condition is to create a demand for products with recycled content (PwRC). More advanced technologies and available government funding to optimise recycling processes have meant PwRC are currently being produced at unprecedented rates. Within a 13-year timeframe, the C&D waste recovery rate has increased from 61% in 2007 to 77% in 2019 [2]. However, the fundamental issue is to balance supply and demand, which is made partly possible by creating and stimulating end-markets for PwRC. If adequately established, such measures can reduce waste disposal rates, improve CE adoption and boost the economy. Hence, PwRC market development has become the agenda of scholars, practitioners and policymakers in various developed and developing countries, including Australia [11], the UK [12], China [13], India [14], France [15], the US and EU (European Union) countries [16]. In Australia, however, despite efforts to establish markets for construction PwRC, its performance is far from ideal [17].

This book therefore aims to provide a better understanding of the operation and capacity of end-markets for C&D waste in Australian jurisdictions, in terms of CE principles and a cradle-to-cradle approach for the sustainable management of C&D waste.

1.2. Content

Firstly, this publication analyses 102 federal, state and territory government policies, standards, guidelines, legislation and specifications to determine the extent to which the Australian regulatory frameworks are in favour of or against C&D waste market development. Then, this study investigates the major barriers to transition towards a CE and several strategies to overcome these barriers are suggested accordingly. Sustainable procurement (SP) is identified as a policy instrument that aims to achieve value for money on a whole life basis in terms of generating benefits not only to the organisation, but also to society and the economy, while minimising damage to the environment. Within this context, a detailed exploration of the application of SP policies and practices at state, national and international levels is conducted. A model is then proposed to facilitate the adoption of SP, based on the research findings.

Next, this book presents the findings from interviews conducted to capture the C&D waste recovery stakeholders' perceptions of COVID-19's impacts, challenges and opportunities regarding C&D waste end-market creation and stimulation in Australia. In total, 27 experts with relevant experience from New South Wales, Victoria, Queensland, South Australia and Western Australia were interviewed between May and July 2021. Drawing on the key findings, a model to optimise the creation and simulation of end-markets for C&D waste in Australia was developed. It maps out how various enablers of market development can be implemented to optimise the interplay among various stakeholders in the C&D waste space. The stakeholders in this model include the government sector (public organisations and policymakers), recyclers, C&D professionals, and sustainability rating system and quality assurance auditors. As proposed in this model, the major improvements should be facilitated by the government sector through various strategies such as SP, raising awareness and financial support.

Lastly, a series of recommendations are provided to transform behaviour in the C&D waste management ecosystem in different states and to increase uptake of PwRC in construction projects.

Conclusion

Valuable insights into five key areas are provided in each chapter (Figure 1), including (1) analysis of national and jurisdictional policy frameworks that affect creation and stimulation of end-markets for waste materials; (2) assessment of the operation of existing end-markets for five strategic construction materials (i.e., brick, concrete, asphalt, steel and glass); (3) exploration of the application of SP policies and practices at state, national and international levels; (4) impacts of COVID-19 on the C&D waste industry; and (5) identification of the main factors affecting C&D waste end-market creation and stimulation, and building a model for waste market development.

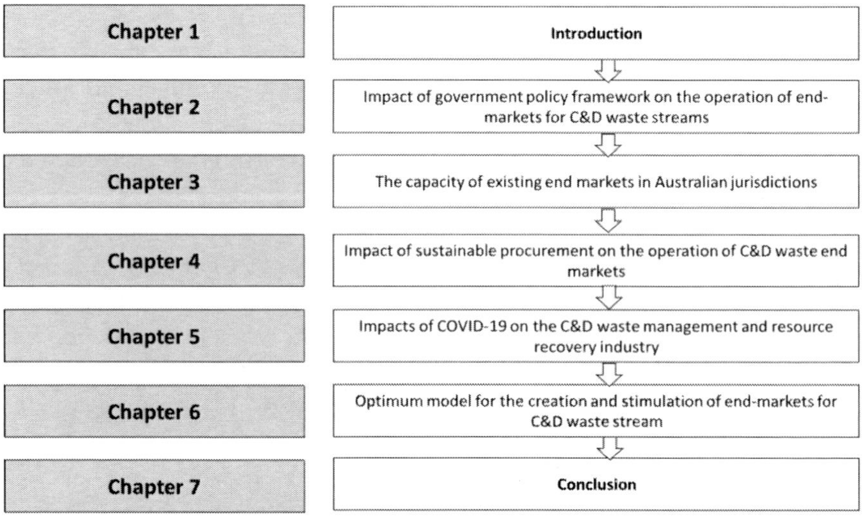

Figure 1. Structure of the book.

Chapter 2

Impact of Government Policy Frameworks on the Operation of End-Markets for C&D Waste Stream

2.1. Introduction

With rapid urbanisation around the world, construction activities continue to expand to an unprecedented extent. As a result, currently the AEC industry accounts for 35% of the total waste sent to landfill worldwide [18]. For instance, in the UK, 49 Mt of C&D waste was generated in 2014; more than half of this amount was landfilled [19]. In China, the annual C&D waste generation is 2.36 BT [20], and in the US, this figure was 516 Mt in 2017 [21]. In Australia, C&D activities have grown substantially in the recent decades, leading to the generation of a large amount of waste. Consequently, C&D waste stream comprise 44% of the total waste generated, reaching 29 Mt annually [2]. The average annual growth of C&D waste generation is currently at 2%, and about 6.7 Mt of this waste stream is landfilled [2]. In response, development of a market for PwRC (including C&D waste) has been identified as a sustainable solution to reduce waste disposal [22]. Hence, there are urgent calls to create a market for C&D waste materials [3].

2.2. Australia

Within the Australian context, according to Australian National Waste Policy [23] the CE can be described through five principles:

> (1) Avoid waste (2) improve resource recovery, (3) increase the use of recycled material and build demand and markets for recycled products, (4) better manage material flows to benefit human health, the environment and the economy and (5) improve information to support innovation, guide investment and enable informed consumer decisions (p. 11)

The third principle highlights creation and stimulation of markets for PwRC. Furthermore, in this policy, Strategy 14 emphasises market development and research (p. 16). Estimations, based on the current solid waste generation rates in Australia, project that Australian recycling capacity must increase by 400% by 2040 to address the issue of solid waste in the future [24]. Several internal and external factors strongly influence the development of a domestic market for waste and recyclables. Among others, a supportive regulatory framework is found to have a significant effect.

Waste regulatory frameworks must be set to be in favour of local market development and the implementation of an effective CE. The issues that must be addressed in this regard are inconsistency in state and territory waste regulations throughout Australia; clarification on the definition of waste including when waste becomes a resource and is not liable for landfill levy; weak or inconsistent regulations that encourage illegal dumping and stockpiling activities; and inconsistent reporting obligations. Hence, this study aims to identify and review jurisdictional regulations, specifications, guidelines and standards affecting the development and operation of end-markets for C&D waste stream. This chapter presents the results of an analysis of waste regulatory frameworks in eight Australian states and territories.

Table 1. Legislation portal web links in different states and territories

State and territory	Portal weblink
Federal Government	https://www.legislation.gov.au/
Australian Capital Territory	https://www.legislation.act.gov.au/
New South Wales	https://www.legislation.nsw.gov.au/#/
Northern Territory	https://legislation.nt.gov.au/
State and territory	Portal weblink
Queensland	https://www.legislation.qld.gov.au/
South Australia	https://www.legislation.sa.gov.au/index.aspx
Tasmania	https://www.legislation.tas.gov.au/
Victoria	https://www.legislation.vic.gov.au/
Western Australia	https://www.legislation.wa.gov.au/

2.3. Approach

The approach consists of a comprehensive review of secondary data (i.e., policies, regulations, guidelines, government reports, technical reports and key journal articles) from publicly available sources. The regulation of waste primarily takes place at the state and territory government level. In each state and territory, as well as federally, a legislation portal makes jurisdictional legislation (acts and subordinate regulations) freely available. Table 1 lists the web addresses of these portals for federal, state and territory governments.

2.4. Results: Review of Regulations, Specifications, Guidelines and Standards

This section of the chapter outlines regulations, guidelines and policies that affect the development and operation of C&D waste end-markets. The review involves the analysis of national and jurisdictional (eight Australian states and territories) documents that are publicly available. Furthermore, public organisations that are responsible for or contribute to developing such end-markets in each state and territory are identified.

2.4.1. Australia

As C&D waste regulation and policy development mostly takes place at the state and territory level, the federal government has set limited legislation for end-market development for PwRC. The Product Stewardship Act 2011 [25] sets criteria for developing stewardship programs in Australia. It provides a distinct definition of waste, which is allocated to a product after it has been disposed of. In the Act, when a product is deemed to be waste, it is still under the product life definition:

> "Life of a product includes: (a) the time when the product begins to be manufactured; and (b) the time when the product is waste" (p. 12)

In terms of policy, the government document that has the most significant impact on end-markets for PwRC is the National Waste Policy 2018 [23]. The Policy was prepared by the Australian government, state and territory governments, and the *Australian Local Government Association*. The Policy guides ongoing collaboration between all Australian governments, business and industries. Through various strategies highlighted in this Policy, it advocates end-markets for PwRC via a common approach, SP by governments, and market development and research. In 2019, the Policy's Action Plan [35] was released to set the targets and actions required to implement the Policy's strategies. The key actions that affect end-market creation and stimulation through governments' purchasing power to increase recycling include stimulating demand for PwRC relative to virgin materials; encouraging innovation and investment in recycling to meet demand from new markets; supporting domestic jobs and industries by retaining the value of PwRC; and encouraging economy-wide behaviour change.

At the national level, efforts to stimulate end-markets for PwRC are typically led by the Federal *Department of Agriculture, Water and the Environment* (DAWE). The Department published the Sustainable Procurement Guideline, which is based on the Commonwealth Resource Management Framework, NWP 2018 and the Sustainable Development Goals. The Guideline maps out the pathway to the government SP approach, including consideration of the purchase of goods and infrastructure that can be re-used, repaired and recycled, and contain PwRC. The Department also published Across the Supply Chain, a guide to analysing the supply chain of C&D waste resources to help develop effective end-markets for materials derived from the C&D waste stream. It provides several case studies across Australia that highlight the sustainable outcomes of using PwRC in construction projects.

2.4.1.1. Regulatory Framework

Table 2 presents a summary of regulatory statements at the federal level related to PwRC market development.

2.4.1.2. Relevant Organisations

The major organisations that contribute most significantly to market creation and stimulations at the national level are the *DAWE* and *Austroads*. *Austroads* is the collective of the Australian and New Zealand transport agencies, representing all levels of government. The organisation work impacts a wide range of agencies including planning, service, infrastructure, health and safety, public health and policing. Specifically, it provides guidelines on the utilisation of PwRC in infrastructure projects such as Guide to Pavement Technology Part 4E: Recycled Materials [26].

Table 2. Australian federal regulatory statements in favour/against PwRC market development

Act and regulations	In favour of market development	Against market development
Product Stewardship Act 2011	Presents criteria for product stewardship programs Provides regulatory context for waste minimisation and increased recycling and re-using activities; some statements include: a) managing waste from products as a resource b) ensuring that products and waste from products are re-used, recycled, recovered, treated and disposed of in a safe, scientific and environmentally sound way Includes the point at which a product is waste in the product life definition Provides a distinct definition of waste, whereby a product is designated as such after it is disposed	No realised or perceived negative impacts
NWP 2018 - Australian Government	Sets 14 strategies to improve waste management The most relevant strategies are: a) Strategy 5 (Common approach): Implement a common approach towards waste policy and regulation, particularly concerning national opportunities to support the development of markets for recycling b) Strategy 8 (Sustainable procurement by governments): All Australian governments consider environmental issues in their approach to goods and infrastructure procurement and promote demand for PwRC c) Strategy 14 (Market development and research): All Australian governments and businesses generate and report information to support creating and maintaining markets for PwRC	No realised or perceived negative impacts
NWP (Action Plan) 2019 [35]	Presents targets and actions to implement the NWP 2018. The relevant targets are: a) Target 4: Significantly increase the use of PwRC by governments and industry Target 7: Make comprehensive, economy-wide and timely data publicly available to support better consumer, investment and policy decisions Proposes actions to stimulate end-markets such as: a) Stimulating demand for PwRC relative to virgin materials b) Encouraging innovation and investment in recycling to meet demand from new markets c) Supporting domestic jobs and industries by retaining the value of PwRC d) Encouraging economy-wide behaviour change	No realised or perceived negative impacts

Table 2. (Continued)

Act and regulations	In favour of market development	Against market development
Sustainable Procurement Guideline - *DAWE*	Applies the government procurement strategies to the Commonwealth Resource Management Framework, NWP 2018 and Sustainable Development Goals Suggests consideration of purchasing goods and infrastructure that can be re-used, repaired and recycled, and contain recycled content	No realised or perceived negative impacts
Guide to Pavement Technology Part 4E: Recycled Materials [26]. - *Austroads*	Presents the latest information and specifications of PwRC to be used in construction projects Suggests a framework for determining the suitability of PwRC in roadworks Determines physical and chemical requirements and relevant testing methods	No realised or perceived negative impacts
Construction and Demolition Waste Guide – Recycling and Re-use Across the Supply Chain - DAWE	Presents information about the supply chain of C&D waste resources Provides case studies in which C&D waste PwRC are used	No realised or perceived negative impacts
Environmental Sustainability Policy - *Department of Services Australia*	Provides suggestions for C&D waste management to the Department offices; these include: a) Comply with the above legislation, regulations and policies b) Apply the environmental standards for waste management as outlined in the Policy c) Re-use and/or recycle—where cost-effective and practical d) Establish re-use and recycling disposal infrastructure for large office relocations to optimise resource recovery e) Maintain accurate and complete records for reporting purposes on the types of disposal infrastructure at each site, the types and units of items being disposed of and the method of disposal	No realised or perceived negative impacts

Table 3. ACT regulatory statements in favour/against PwRC market development

Acts and regulations	In favour of market development	Against market development
Environment Protection Act 1997 *ACT EPA*	• One of the Act's objects is to promote waste minimisation and the re-use and recycling of materials	• No clear definition for C&D waste • Waste is always waste regardless of its value
Waste Management and Resource Recovery Act 2016 -*Transport Canberra and City Services Directorate*	• This Act implies support for market development through the following objects: a) Manage waste through maximising the recovery and re-use of resources b) Support innovation and investment in waste management c) Promote responsibility for waste reduction	• No clear definition for C&D waste • Waste is always waste regardless of its value • Proximity principle applied to waste materials
Environment Protection Regulation 2005 - *ACT EPA*	• Not evident	• Waste is always waste regardless of its value • Clean fill is not excluded from waste category
Waste Management and Resource Recovery Regulation 2017 - *Transport Canberra and City Services Directorate*	• Not evident	• No definition for C&D waste, no list of recyclable C&D waste

2.4.2. Australian Capital Territory (ACT)

This section analyses the regulatory framework that governs and guides market creation and stimulation activities in the ACT. Included in the review are strategies, standards, legislation and specifications. The primary pieces of legislation in the ACT are the Waste Management and Resource Recovery Act 2016 (WMRR Act) and the Environment Protection Act 1997. The primary pieces of legislation do not account for the end-of-life options and waste is always considered as waste regardless of its value. However, in the Development Control Code [27] a set of mandatory requirements are specified, which state that development projects with more than 20 m^3 of waste generation must recycle and re-use C&D waste materials. The ACT has no specifications for the application of C&D waste PwRC in infrastructure projects. The leading ACT organisations that are responsible for the creation and stimulation of the C&D waste market are the *ACT EPA, Transport Canberra* and the *City Services Directorate*.

2.4.2.1. Primary Legislation

The two primary pieces of legislation are the WMRR Act and the Environment Protection Act 1997. Their subordinate regulations do not provide significant support for C&D waste market creation and stimulation. The major issues are related to having no clear definition of C&D waste, which might create confusion among waste businesses and operators, leading them not to account for waste end-of-life options. Table 3 provides a summary of regulatory statements enshrined in ACT legislation in favour of and against market development.

2.4.2.2. Guidelines, Policies and Strategies

This section analyses policies and strategies affecting market development in the ACT. In 1996, the ACT was the first jurisdiction in the world to employ a 'No Waste' strategy. Since its implementation, this strategy has set the waste policy for four governments (1996–2004). The cornerstone of this strategy was based on the notion that waste material is a resource, not waste. Following this strategy, three other policies and strategies were developed that have guided waste management planning and activity in the territory. The ACT's Waste Strategy (2015–2025) is a pioneering document that provides practical strategies that are relevant to efforts to create and stimulate PwRC end-markets. The Waste-to-Energy Policy 2020–25 [28] recommends SP to create a market for PwRC. The major change reflected in that policy is a pause on the new plans for energy recovery from waste activities throughout the territory as of 2020. It also limits the market availability for combustible C&D waste materials such as timber. The 2015 Sustainable Procurement Policy is among the first policies published in the Australian states and territories focusing on the cradle-to-cradle approach, which will have multiple sustainable (environment and economic) outcomes. Among the specified outcomes, end-of-life options and job creation through market development are the most relevant considerations. However, this policy does not provide any specific information on the use of C&D waste PwRC by the *ACT Government*. Table 4 provides a summary of the strategies and factors that can trigger market development.

Table 4. ACT guidelines and strategies in favour/against PwRC market development

Guideline/strategy	In favour of market development	Against market development
ACT Waste Management Strategy 2011–2025 - Environment, Planning and Sustainable Development Directorate)	Advises the following strategies: a) Develop markets for recyclable materials and strengthen the regional connection b) Disincentivise landfill including appropriate pricing and regulation c) Government sustainable procurement d) Encourage on-site re-use for C&D waste	No realised incentive for C&D waste market development
ACT Waste-to-Energy Policy 2020–25 - ACT Government	Government seeks to create markets for PwRC through government procurement It rules out energy from recovery activities throughout the territory	No realised incentive for C&D waste market development
Sustainable Procurement Policy 2015 - ACT Government	The policy supports the cradle-to-cradle approach and suggests several sustainable impacts to consider when procuring goods and services: a) Reusability and/or recyclability b) Product efficiency and longevity: options for re-use, repair, upgrade or modification to increase product life c) Recycled content of goods The policy indicates the outcome of sustainable procurement as follows: a) Job creation (e.g., green technologies, use of local suppliers, creating markets for PwRC, back to work schemes) b) end-of-life options (e.g., recyclability, resource recovery)	No specific mention of C&D waste PwRC

2.4.2.3. Standards, Codes and Specifications

Among all the government-issued documents, the Development Control Code [27] is the most relevant source of information about the C&D waste stream in the ACT. This document lists the obligatory requirements for re-using and using PwRC in development projects. These requirements are provided in Table 5. The ACT has not published any specifications with regards to the use of PwRC in construction and infrastructure projects.

Table 5. ACT government document statements in favour/against PwRC market development

Standard/code/specification	In favour of market development	Against market development
Development Control Code for Best Practice Waste Management in the ACT 2019 – Transport Canberra and City Services Directorate	The general waste definition excludes recyclable materials C&D waste definition is provided Best practice waste management for C&D waste provided through *Part 6. Mandatory requirements- demolition, excavation and construction* with the following conditions for development activities that generate more than 20 m^3 of waste: a) Minimum 90% of all demolition and excavated waste generated on development must be re-used or recycled or both b) Minimum 5% of all construction waste generated on development must be re-used or recycled	No realised incentive for C&D waste market development

2.4.2.4. Relevant Organisations

The three major organisations that contribute most significantly to market creation and stimulations are the *ACT EPA*, the *Environment, Planning and Sustainable Development Directorate*, and *Transport Canberra* and *City Services Directorate*.

2.4.3. New South Wales (NSW)

This section analyses the regulatory framework that governs and guides market creation and stimulation activities in NSW. Included in the review are strategies, standards, legislation and specifications. The primary piece of legislation is the Protection of the Environment Operations (Waste)

Regulation 2014. The state's SP policy prioritises the use of PwRC in government projects. The policy advocates procuring goods and services that are eco-labelled (e.g., with environmental declarations), compliant with environmental-based standards, and registered in recognition programs and systems (e.g., sustainable choice, sustainability advantage and recycled content). The *NSW Environment Protection Authority (NSW's EPA)* and *Transport for NSW* (*TfN*) are the main organisations that contribute to creating and stimulating C&D waste material end-markets in NSW. The EPA provides leadership to ensure NSW has a fair, modern and well-regulated waste industry. This organisation provides several C&D waste resource-specific exemption documents that outline how the consumers of PwRC are exempted from waste levies across the state.

2.4.3.1. Primary Legislation

The definition of waste in the Protection of the Environment Operations Act 1997 is extended to the end of waste (EoW) stage of materials irrespective of it being processed, recycled, reused or recovered. The Protection of the Environment Operations (Waste) Regulation 2014 is considered to be the most significant piece of legislation regarding waste in NSW. It provides exemptions for waste and its derivatives for reclaiming and depositing on the land or for use as a fuel. The implication of this regulation is further extended through several order and exemption documents published by *NSW's EPA*. Table 6 provides a summary of regulatory statements enshrined in NSW legislation in favour of and against market development.

2.4.3.2. Guidelines, Policies and Strategies

NSW's SP guideline is the most comprehensive document on this topic in Australia. It maps out the state's foundations and strategy for SP of goods and services. According to this document, prioritised for procurement are those materials that are eco-labelled (i.e., with environmental declarations) and/or are registered in national and state recognition programs and systems (e.g., sustainable choice, sustainability advantage and recycled content). In practice, NSW's SP follows the purchase strategy, understanding need and the market, developing a plan and preparing a specification, assessing responses and supplier selection, and reviewing and identifying improvement. Further, *TfN* has released a technical guide for using and re-using various waste streams including C&D waste materials in road construction activities. Besides outlining waste management best practices, it explains the pathway to obtain

Table 6. NSW regulatory statements in favour/against PwRC market development

Act and regulations	In favour of market development	Against market development
Protection of the Environment Operations Act 1997 – *NSW EPA*	Includes C&D waste in the litter definition Defines waste owner Prescribes the local council responsibility in developing and administrating a waste reduction and environmental sustainability scheme Promotes the reduction in the use of materials and the re-use, recovery or recycling of materials	Waste is always waste regardless of its intended fate and status Resource recovery excludes waste energy recovery
Waste Avoidance and Resource Recovery Act 2001 – *NSW EPA*	Adopts waste hierarchy The objectives of the Act that relate to C&D waste end-markets are: a) To encourage the most efficient use of resources b) To ensure that industry shares with the community the responsibility for reducing and dealing with waste c) To ensure the efficient funding of waste and resource management planning, programs and service delivery d) To achieve integrated waste and resource management planning, programs and service delivery on a state-wide basis Considers environmental policies such as an extended producer responsibility scheme and product stewardship Describes the EPA priorities	Waste is always waste regardless of its intended fate and status No specific statement related to C&D waste end-markets
Protection of the Environment Administration Act 1991 – *NSW EPA*	Prescribes the NSW EPA responsibilities and objectives; some of its objectives are: a) Encouraging the re-use and recycling of materials and encouraging material recovery b) Setting mandatory targets for environmental improvement c) Environmental goals, having been established, should be pursued most cost-effectively by establishing incentive structures including market mechanisms that enable those best placed to maximise benefits or minimise costs to develop their solutions and responses to environmental problems	No specific statement related to C&D waste end-markets
Protection of the Environment Operations (Waste) Regulation 2014 – *NSW EPA*	Provides exemptions for waste and its derivatives for reclaiming and depositing on the land or for use as a fuel	No specific statement related to C&D waste end-markets

Table 7. NSW guidelines and strategies in favour/against C&D waste PwRC market development

Guideline/strategy/policy	In favcur of market development	Against market development
Waste Levy Guidelines 2018 – *NSW EPA*	• Defines C&D waste stream • Outlines the waste materials and relevant exemption requirements to be used on roads, other construction works and landfill facilities as bedding layers	No realised or perceived negative impacts
NSW Waste Avoidance and Resource Recovery Strategy 2014–21 – *NSW EPA*	• Describes the state's funding strategy (Waste Less, Recycle More) for developing new markets for PwRC • Advocates waste management strategies such as product stewardship, industrial symbiosis and sustainable procurement • Outlines the responsibility of public departments and organisations to manage waste	No realised or perceived negative impacts
Sustainable Procurement Guide 2017 – *NSW Government*	• Maps out the state's foundations (e.g., the quadruple bottom line, whole-of-life costs, lifecycle assessment, circular economy, and community and social impact) and strategy for sustainable procurement of goods • Outlines the state's procurement reliance on eco-labelling (e.g., environmental declarations), standards, and recognition programs and systems (e.g., sustainable choice, sustainability advantage and recycled content) when comparing services and goods	No specific statement related to C&D waste materials
Environmental Guidelines Solid waste landfills 2016 -*NSW EPA*	Specifies some of the C&D waste materials (e.g., crushed building waste and glass) that can be used for daily cover and final capping in landfill sites	No realised or perceived negative impacts
Management of Wastes on Roads and Maritime Services Land 2014 – *TfN*	• Sets out the *TfN* approval and waste management procedures for utilising *TfN* land sites for road construction activities • Explains the procedure to obtain an exemption for certain C&D waste materials	No realised or perceived negative impacts
Technical Guide Management of Road Construction and Maintenance Wastes 2016 – *TfN*	Provides information on the statutory and *TfN* policy requirements for the management of wastes associated with road construction and maintenance works	No realised or perceived negative impacts

Table 7. (Continued)

Guideline/strategy/policy	In favour of market development	Against market development
Waste Classification Guidelines Part 1: Classifying Waste- *NSW EPA*	• Provides a clear definition of C&D waste, uses "unsegregated" as an inclusion criterion • Excludes excavated soil from the C&D waste definition	Not clear whether a segregated C&D waste is a source of waste
NSW Energy from Waste Policy Statement 2015v [29] – *NSW EPA*	• The policy statement objectives are: a) Promote the source separation of waste where technically and economically achievable b) Drive the use of best practice material recovery processes Ensure only the residual from bonafide resource recovery operations are eligible for use as a feedstock for an energy recovery facility • Lists eligible waste fuels, which includes uncontaminated wood waste • Describes the landfill levy exemptions that apply to waste fuels and outlines the requirements of using residual waste from mixed C&D waste processing facility	Treated timber is excluded from the eligible waste fuels list
Resource Recovery Orders and Exemptions- *NSW EPA*	Provides a brief guideline on EPA resource recovery orders and exemptions objectives and requirements	No realised or perceived negative impacts

Table 8. NSW government document statements in favour/against PwRC market development

Standard/code/specification	In favour of market development	Against market development
NSW Extended Producer Responsibility Priority Statement 2010 – *Department of Environment, Energy and Science*	From the C&D waste stream, only treated timber, PVC and packaging are listed in the scope of the document	No specific statement related to the C&D waste stream
Specification for Supply of Recycled Material for Pavements, Earthworks and Drainage 2010 – *IPWEA (NSW)*	Provides a specification for the supply of PwRC, primarily crushed concrete, brick and reclaimed asphalt blends for the following uses: a) Road base for roads with light to medium traffic b) Bedding material for paving blocks to be used in pedestrian areas, car parks, shopping centres, etc. c) Select fill for use on subgrades to enhance strength or for raising site levels, particularly in roadways and beneath buildings d) Drainage medium for drainage lines and drainage structures	No realised or perceived negative impacts
NSW's EPA: Recovered aggregate Recovered railway ballast Recovered glass sand Reclaimed asphalt pavement Cement fibreboard Excavated natural material Excavated public road material Plasterboard	• Provides chemical, other material requirements and testing procedures for exemption purposes • Describes applications	No realised or perceived negative impacts

EPA approval/exemption to serve such purposes. Lastly, the NSW Energy from Waste Policy Statement 2015 [29] encourages this waste management method by providing levy exemptions for uncontaminated wood and residual waste from mixed C&D waste processing facilities. Table 7 presents a summary of guidelines, policies and strategies related to C&D waste market development in NSW.

2.4.3.3. Standards, Codes and Specifications

NSW's Extended Producer Responsibility Priority Statement 2010 lists some C&D waste materials that fall under current product stewardship programs. These programs aim to share the responsibility of the EoW stage with their suppliers/manufacturers. In 2014, the NSW EPA published several Resource Recovery Orders/Exemptions under Part 9, Clause 93 of the Protection of the Environment Operations (Waste) Regulation 2014. The orders impose the requirements that must be met by suppliers of C&D waste resources to which the corresponding exemptions apply. These documents are effective enablers towards stimulating end-markets for C&D waste resources across the state. Additionally, the Institute of the Public Works Engineering Australia has specified the use of C&D PwRC in infrastructure projects. The specification is considered at the national level. Table 8 presents a summary of standards, codes and specifications related to C&D waste market development in NSW.

2.4.3.4. Relevant Organisations

The *NSW EPA* and *TfN* are the main government agencies that contribute to creating C&D waste material end-markets in NSW. The *EPA* provides leadership to ensure NSW has a fair, modern and well-regulated waste industry. *TfN*'s Environmental Policy provides the framework for ongoing improvements in *TfN*'s environmental performance. A relevant objective in this framework is to promote the efficient use, re-use and recycling of resources, and the minimisation of waste. To promote the management of waste and re-use of resources in the relevant construction activities, *TfN* published a technical guide. Meanwhile, *NSW's EPA* has published several Resource Recovery Orders/Exemptions for multiple C&D waste resources that contribute to developing and stimulating relevant end-markets.

2.4.4. Northern Territory (NT)

This section analyses the regulatory framework that governs and guides market creation and stimulation activities in the NT. Included in the review are strategies, standards, legislation and specifications. The NT waste regulatory framework builds on the Waste Management and Pollution Control Act 1998. The leading government agencies that provide support for market creation and stimulation are the *Department of Environment and Natural Resources (DENR)*, the *Department of Infrastructure, Planning and Logistics (DIPL)*, and the *Northern Territory Environment Protection Authority (NT's EPA)*. The Department prepares the main pieces of legislation. The NT does not have a sustainable procurement policy with a focus on government commitment to procure PwRC; however, in its Procurement Governance Policy 2019, the sustainable outcomes are emphasised. Furthermore, the application of C&D waste PwRC is encouraged in two standard specifications developed by *DIPL*.

2.4.4.1. Primary Legislation

There are four pieces of primary legislation in the NT that form the waste regulatory framework. The Department of Environment and Natural Resources administers these acts and regulations. Aside from the exemption for using waste materials in land reclamation activities with environmental protection approval as specified in the 1998 Act, the primary legislation provides little support towards C&D waste market creation and stimulation through general sustainability principles. Overall, the government agencies consider waste to always be waste, irrespective of its value. Table 9 provides a summary of NT regulatory statements in favour of and against market development.

2.4.4.2. Guidelines, Policies and Strategies

Except for the Waste Management Strategy document, the relevant government departments and agencies have not published any guidelines or policies about C&D waste market development in the NT. As summarised in Table 10, the NT's waste strategy encourages three actions that will boost market development. However, this strategy does not indicate any specific actions for the C&D waste market.

Table 9. NT regulatory statements in favour/against PwRC market development

Act and regulations	In favour of market development	Against market development
Waste Management and Pollution Control Act 1998 - *DENR*	The Act exempts construction activities from environmental protection approval when waste rock, rubble and other inert materials are used for land reclamation purposes	• No clear definition for C&D waste • Waste is always waste regardless of its value
Environment Protection Act 2019 - *DENR*	The Act promotes a waste hierarchy that prefers waste recycling and re-using over the disposal	• No clear definition or information about C&D waste • No encouragement for sustainable government procurement
Waste Management and Pollution Control (Administration) Regulations 1998 - *DENR*	No evident support	• No clear definition or information about C&D waste • No encouragement for sustainable government procurement
Environment Protection Regulations 2020 - *DENR*	No evident support	• No clear definition or information about C&D waste • No encouragement for sustainable government procurement

Table 10. NT guidelines and strategies in favour/against PwRC market development

Guideline/strategy	In favour of market development	Against market development
Waste Management Strategy for The Northern Territory 2015–2022 – *NT EPA*	Under the management actions, it advises facilitating opportunities to connect waste recovery and re-use markets with key waste producers. Other measures recommended are: a) Collaborate with the Waste and Recycling Industry of the NT (WRINT) to identify emerging trends and issues requiring multi-faceted solutions; b) Work with industry and government agencies to demonstrate the economic incentives available through improved waste management and resource recovery	• No specific action/strategy for C&D waste market creation and stimulation • Does not support sustainable government procurement

Table 11. NT government document statements in favour/against PwRC market development

Standard/code/specification	In favour of market development	Against market development
Standard specification for Environmental management V.2 - *DIPL*	• Implement measures to reduce, re-use and recycle waste products/materials including soil, road pavement materials and concrete • Demonstrate the priority use of materials and products that maximise the use of PwRC wherever these are cost- and performance-competitive and are at least the environmental equivalent of the non-recycled alternative	Only general recommendations are provided
Standard Specification for Roadworks - *DIPL*	• Source the material from some sources including window (or flat) glass and plain ceramic. The application should comply with specifications enshrined in another document. [30] • Addition of crush and screen reclaimed asphalt pavement no more than 15% (by mass) in base layers or 10% in the wearing course	Other potential PwRC are not specified

2.4.4.3. Standards, Codes and Specifications

The *DIPL* has provided standard specifications for re-using PwRC in government projects. These two standards encourage using C&D waste materials in construction projects. In the environmental management standard specification, the three conditions in which PwRC are preferred are: (1) when they perform better in terms of cost; (2) when they perform better in terms of technical performance; or (3) when they are the environmental equivalent of their fresh materials alternatives. In the roadworks standard specification, only the application of recovered crushed glass (RCG) is advised. Table 11 presents a summary of standards, codes and specifications related to C&D waste market development in the NT.

2.4.4.4. Relevant Organisations

The *DENR*, *NT EPA* and the *DIPL* are the government agencies that address of issues related to C&D waste market development in the NT. The *DENR* administers the main pieces of legislation and the *DIPL* provides specifications for using PwRC in construction projects. The *NT EPA* is the agency that developed the territory's waste strategy.

2.4.5. Queensland (Qld)

Qld is the first state in Australia that has separated waste from a recoverable resource in its regulatory framework. The legislative framework effectively facilitates waste management and resource recovery stakeholders' engagement through several consultation opportunities throughout the development of legislative documents. The main public organisations that contribute to C&D waste materials end-markets development in Qld are the *Department of Environment and Science (DES)*, the *Department of State Development, Tourism, and Innovation (DSDTI)* and the *Department of Transport and Main Roads (DTMR)*. The state has specified the EoW framework in the Waste Reduction and Recycling Act 2011, which promotes resource recovery opportunities and aims to transform the perception of waste from being seen as waste to being valued as a resource. A waste producer may supply waste as a resource under an EoW code provided they have registered with the *DSE* and can comply with the requirements of the code. *DTMR* has developed two specifications for using recycled glass aggregates and reclaimed asphalt pavement (RAP).

2.4.5.1. Primary Legislation

The Waste Reduction and Recycling Act 2011 (current as of 1 July 2019), detailed in 16 chapters, is the most updated waste act in Australia that covers various waste management strategies and enablers. The strategies include the polluter pays principle, the user pays principle and product stewardship. Included in enablers are discounted landfill levy rates for residue waste generated by recycling activities and a disposal ban. These enablers stimulate the market for C&D waste-derived products. The Act also allows for pubic and industry submissions for developing an EoW code; this aims to assist in meeting stakeholders' expectations and further stimulate a market for C&D waste. It obligates the local governments to prepare a specific waste reduction and recycling plan that follows the Act, its contextual conditions, and waste management and resource recovery principles and consultation with the public. Table 12 provides a summary of regulatory statements enshrined in Qld legislation in favour of and against market development.

2.4.5.2. Guidelines, Policies and Strategies

The Queensland Waste Avoidance and Resource Productivity Strategy (2014–2024), developed by the DES, categorises some C&D waste materials with high volume and current resource value as the state's priority waste. Furthermore, this document proposes several priorities for Qld with regards to end-market creation and stimulation. These priorities are classified under four major objectives: Driving cultural change; avoidance and minimisation; re-use, recovery and recycling; and management, treatment and disposal. The *DES* also provides an EoW guideline, as per the WRR Act requirements, that prescribes the process for people who are using or intending to re-use waste as a resource. A waste producer may supply waste as a resource under an EoW code provided they have registered with the DES and can comply with the requirements (i.e., specified criteria and quality characteristics) of the code. If waste material is approved as a resource under the EoW framework, it is no longer considered waste. The Qld Government Procurement Policy 2019 [31] aims to manage demand and reduce waste and manage the consumption of valuable resources. However, the use of PwRC is not recommended. Table 13 presents a summary of guidelines and strategies related to Qld's C&D waste market development.

Table 12. Qld regulatory statements in favour/against PwRC market development

Act and regulations	In favour of market development	Against market development
Environmental Protection Act 1994 - *DES*	Defines waste in a detailed approach as follows: • As anything, other than an end of waste (EoW) resource, that is a) leftover or an unwanted by-product, from an industrial, commercial, domestic or other activity; or b) surplus to the industrial, commercial, domestic or other activity generating the waste c) waste can be a gas, liquid, solid or energy, or a combination of any of them d) A thing can be waste whether or not it is of value • Excludes C&D waste from the definition of landfill/disposing of waste	No realised or perceived negative impacts
Waste Reduction and Recycling Act 2011 - *DES*	• Defines C&D waste and adopts a waste hierarchy model • Introduces a few different waste management strategies: i.e., the polluter pays principle, user-pays principle and product stewardship • Considers disposal ban waste • Recommends discounted landfill levy rates for residue waste in recycling activities • Considers a non-levy zone across the state • If waste used as feedstock for a recycling activity was generated outside the non-levy zone, all of the residue waste generated by the recycling activity is taken to be generated outside the non-levy zone • Requires local governments to consider the following when adopting a waste reduction and recycling plan: a) Population profile b) Residential, industrial and commercial development c) Amounts and types of waste generated d) Services, markets and facilities relevant to dealing with different types and amounts of waste	

Act and regulations	In favour of market development	Against market development
Waste Reduction and Recycling Act 2011 - *DES*	e) The waste and resource management hierarchy f) The waste and resource management principles g) Achieving the state's waste management strategy objectives h) Consulting with the public about the plan and amendments • Details the requirements for waste management and resource recovery data reporting • Prescribes the end of waste (EoW) when it becomes a resource • Requires state government agencies to prepare and publish a Waste Management Plan and to report annually against the progress of the Plan to the Department of Environment and Science	• The obligation of local government for recycling activities might induce inconsistencies in efforts to implement a uniform management system across the state • Proximity principles may limit market development
Environmental Protection Regulation 2008 - *DES*	C&D waste is categorised as general waste	• No specific statement about the C&D waste stream • No statement directly related to PwRC market development
Waste Reduction and Recycling Regulation 2011	Defines C&D waste	• No specific statement about the C&D waste stream • Market development enablers are not included

Table 13. Qld guidelines and strategies in favour/against PwRC market development

Guideline/strategy	In favour of market development	Against market development
Waste–Everyone's responsibility Queensland Waste Avoidance and Resource Productivity Strategy (2014–2024) - *DES*	Classifies high volume C&D wastes with a current resource value (i.e., concrete, treated timber and plasterboard) as the state's priority waste Proposes four main objectives for the state's waste management: Driving cultural change; avoidance and minimisation; re-use, recovery and recycling; and management, treatment and disposal. The following priorities under these objectives are related to C&D waste end-markets: a) Awareness and communication b) Roles and responsibilities c) Education and training d) Industrial ecology and infrastructure planning for driving change e) R&D f) Improved data to highlight business opportunities g) Regional collaboration and partnerships h) Product stewardship i) Market development and appropriate incentives j) Appropriate regulation and enforcement k) Full cost accounting of all disposal	No realised or perceived negative impacts
Waste Management and Resource Recovery Strategy - *DES*	Includes several strategies to develop a market for recyclable materials: a) Landfill disposal bans on selected waste streams b) Explore product stewardship schemes c) Develop a policy for energy from waste d) Assess the opportunities of the circular economy model for Qld e) Collect and amalgamate data to understand f) Material flows across the economy and address knowledge gaps g) Continuously improve and reform waste-related legislative frameworks h) Support the commercialisation of successful recycling and re-manufacturing technologies	No realised or perceived negative impacts

Guideline/strategy	In favour of market development	Against market development
End of Waste (EoW) Guideline – *DES*	Prescribes the process for people who are using or intending to re-use waste as a resource such as how to become a registered resource producer In preparing an application to use the benefits of the EoW framework, the applicant has to consider if a new market for the given waste is likely to be developed	A resource should be comparable to any virgin material for which it would substitute
Resource Recovery Industries 10 Year Roadmap and Action Plan 2019 - *DSDTI*	The action plan specifies core waste types in Qld including built environment waste Amount the four main strategies, Strategy 2 focuses on market and supply chain development. The relevant actions are below: a) Deliver a comprehensive analysis of the resource recovery market sector b) Develop industry attraction programs focused on technologies and industries that will utilise recycled output to develop secondary markets c) Investigate and report on export opportunities for value-added PwRC for the development of secondary markets d) Coordinate supply chain development services and activities across the state, including customised activities to meet local and regional needs, where appropriate e) Consult with regional teams to understand issues that relate to waste management and resource recovery specific to their region f) Deliver business development and capability programs to enable SMEs and small regional councils to participate fully in market opportunities g) Analyse aggregated data from the RRIDP to inform ongoing market development initiatives h) Promote exemplar resource recovery precincts in Qld to demonstrate investment readiness i) Review resource recovery initiatives on large government-led projects to develop best practice guidelines j) Investigate opportunities for PwRC to be included in government procurement policies	No realised or perceived negative impacts

Table 13. (Continued)

Guideline/strategy	In favour of market development	Against market development
Waste Reduction and Recycling Plan 2018–2024 - *DES*	Build on the learnings from waste plans of those former agencies and entities comprising *DES* Establish key roles and networks Map major departmental waste streams Empower all staff to engagement in appropriate waste management behaviours	No focus on market development for C&D waste

Table 14. Qld government document statements in favour/against PwRC market development

Standard/code/specification	In favour of market development	Against market development
End of Waste Code Liquid Concrete Washout (ENEW07602719) - *DES*	States when the liquid concrete washout becomes a resource and any relevant requirements and/or conditions for its use	No realised or perceived negative impacts
End of Waste Code Returned Concrete (ENEW07278517) - *DES*	States when the returned concrete becomes a resource and any relevant requirements and/or conditions for its use	No realised or perceived negative impacts
End of Waste Code Solid Concrete Washout (ENEW07602819) - *DES*	States when the solid concrete washout becomes a resource and any relevant requirements and/or conditions for its use	No realised or perceived negative impacts
Transport and Main Roads Specifications MRTS36 Recycled Glass Aggregate - *DTMR*	Sets out the requirements for recycled glass aggregate used in asphalt and unbound granular road pavements, not other applications	No realised or perceived negative impacts
Transport and Main Roads Specifications MRTS102 RAP Material - *DTMR*	Sets out the requirements for RAP material that is used in asphalt	No realised or perceived negative impacts

2.4.5.3. Standards, Codes and Specifications

The Queensland *DES* has released several EoW codes for various waste materials. Currently, there are three specific C&D waste sub-streams, i.e., liquid washout, returned concrete and solid washout. At the time of writing, *DES* is preparing an EoW code for recycled aggregate that is an enabler for C&D waste end-markets development. The Queensland *DTMR* is responsible for developing specifications for using PwRC in infrastructure projects. Currently, the Department has issued two specifications for two C&D waste materials: recycled glass aggregates and reclaimed asphalt. Table 14 presents a summary of standards, codes and specifications related to C&D waste market development in Qld.

2.4.5.4. Relevant Organisations

The main public organisations that contribute to the C&D development of waste material end-markets are the *DES*, the *DSDTI* and the *DTMR*. The *DES* administers the legislation across the state and provides the EoW codes and guidelines and waste strategies. At the time of writing, the *DSDTI* is developing the Resource Recovery Industries 10 Year Roadmap and Action Plan. The *DTMR* developed two specifications for using glass aggregates and RAP material.

2.4.6. South Australia (SA)

SA is a leading state in the operation of an effective C&D waste management system, with a state recycling target set at 90% in 2020 [7]. The primary act in this state, the Environmental Protection Act 1993, excludes recovered resources from the definition of waste resources. SA is among the first of the states and territories to put a SP policy in place for government procurement practices. *Green Industries SA (GISA)*, the SA *Department for Infrastructure and Transport (DIT)* and the *SA Environmental Protection Authority (SA's EPA)* are the government agencies that deal with the C&D waste issues and market development. *GISA* offers market development research grants to stimulate an increase in the quality and market demand for recyclable materials and PwRC. In 2013, the *SA's EPA* released a standard for using waste materials as waste-derived fill (WDF); the standard describes the processes and approval requirements of the application of the C&D waste stream in development activities. Overall, the regulatory framework in this state is in favour of a CE and market development.

2.4.6.1. Primary Legislation

The primary piece of legislation in SA is the Environment Protection Act 1993, which has a specific focus on CE and market development. The Act allows for an exemption of meeting waste management requirements for specific materials as approved recovered resource. The Act objectives promote resource recovery through programs that encourage the industry, public authorities and the community to engage in resource recovery. The subordinate regulation (Environment Protection Regulations 2009) defines some C&D waste materials (e.g., rock, concrete, clay and soil) as clean fill, which is excluded from the levy fee. Table 15 provides a summary of regulatory statements enshrined in SA legislation in favour of and against market development.

2.4.6.2. Guidelines, Policies and Strategies

Several guidelines and strategies facilitate resource recovery and PwRC market development in SA. As already noted, clean waste is exempted from the landfill levy. Meanwhile, *GISA* provides market development with a research grant and infrastructure investment loan scheme to stimulate market demand for recyclable materials and streamline greater local re-manufacturing/reprocessing and adoption of new technologies to reduce contamination level and improve PwRC quality. Notably, the *SA EPA*'s Environment Protection (Waste to Resources) Policy 2010 requires the Authority not to recommend the Minister to ban landfilling for certain waste materials unless there are existing markets for them. The *GISA's* Waste Strategy 2015-2020 fully supports the development of the market for PwRC; the agency is in the process of updating the state waste strategy. Another state policy is the government procurement policy, the Sustainable Procurement Guideline 2012, which stimulates market development through applying demand management strategies such as encouraging the use of second-hand materials. Further, the SA *DIT* has published several guidelines for using clean waste including soil in infrastructure and transport projects; however, the permissible percentage of their usage is kept at a minimum (e.g., 20% by mass for concrete). SA's Waste and Resource Recovery Infrastructure Plan highlights the role of the market as a soft infrastructure in the use of resource recovery in the construction of new infrastructure and provides several actions and strategies to create PwRC markets in the state. Table 16 presents a summary of guidelines and strategies related to C&D waste market development In SA.

Table 15. SA regulatory statements in favour/against PwRC market development

Act and regulations	In favour of market development	Against market development
Environment Protection Act 1993 - *SA's EPA*	Waste management hierarchy is adopted Some materials can be approved as recovered resources and exempt from meeting requirements that apply to waste material The Act objectives include: a) Encourage and assist action by industry, public authorities and the community aimed at pollution prevention, clean production and technologies and resource recovery; b) Promote the circulation of materials through the waste management process and support a substantial market for the recovered resource by: i. programs to encourage and assist the industry, public authorities and the community to engage in resource recovery ii. regulating resource recovery	Waste is always waste regardless of its value and fate—however, approved recovered resource is excluded from this definition C&D waste is not defined
Environment Protection Regulations 2009 - *SA's EPA*	Clean fill is excluded from the waste category and the levy is no payable on clean fill Clean fill includes some C&D waste materials such as clay, concrete, rock, sand and soil Defines waste fill (clean fill) and specifies its chemical and physical properties	Brick waste is not considered a clean fill C&D waste is not defined
Green Industries SA Act 2004	Tasks Green Industries SA to develop the waste strategy and identify targets for resource recovery Lists the functions of *GISA* concerning resource recovery including market development Builds on circular economy principles, the waste management hierarchy, ecologically sustainable development and other best-practice methods	C&D waste is not defined

Table 16. SA guidelines and strategies in favour/against PwRC market development

Guideline/strategy	In favour of market development	Against market development
Environment Protection (Waste to Resources) Policy 2010 - *SA EPA*	The policy required the Authority not to recommend the Minister to declare disposal of waste unless: a) There are existing or developing markets to enable resource recovery b) There are reasonable and practicable existing and new processes, technologies or systems that enable resource recovery	No realised incentive for C&D waste market development
Waste Levy Regulations Guidelines 2016 - *SA EPA*	Levy fee is not appliable to clean fill	No realised or perceived negative impacts
South Australia Waste Strategy 2015–2020 - *GISA*	Some objectives of this strategy include: a) A stable market for investors through a clear policy framework providing a reliable platform for investment decisions b) A resource-efficient economy where the best or full value is secured from products and materials produced, consumed and recovered across the state c) A stable and efficient market for investors; essentially, a clearly articulated policy framework that gives a solid platform for investment decisions The strategy recommends the following market-based tools: a) Policy for setting landfill levies, advance disposal and recycling fees, and deposit-refund and subsidy schemes b) Incentives, grants and loans to promote change and efficiency, enhance environmental performance and innovation, and reward desired behaviours c) Producer responsibility measures including future recycling, costs in purchase prices, influencing buying decisions and supporting the effective recovery of valuable resources	No realised or perceived negative impacts

Guideline/strategy	In favour of market development	Against market development
	Some suggestions specific to C&D waste stream include: a) Encourage the use of secondary materials such as concrete aggregate b) Encourage salvaging and re-use of building materials c) Promote the procurement of re-manufactured material and products	
Sustainable Procurement Guideline 2012 - *SA Government*	• Fosters a viable Australian and New Zealand market for sustainable products and services by supporting businesses and industry groups that demonstrate innovation in sustainability • Stimulates markets through applying demand management strategies such as: a) Re-using, refurbishing or reconditioning the product to extend its life b) Acquiring second-hand or used items	No clear instruction for using C&D waste PwRC
Business Plan 2018-19 - *GISA*	• Suggests a market development research grant that stimulates an increase in the quality and market demand for recyclable materials and PwRC. • Focuses on creating market opportunities for new, sustainable products made from PwRC; expanded market-related activities for existing recycled-content products; improved quality and supply of waste feedstock used in manufacturing recycled-content products; and improved market confidence in PwRC and recycled-content products • Outlines an infrastructure investment loan scheme to provide greater local re-manufacturing/reprocessing and the adoption of new technology to improve quality and reduce contamination level for recyclable materials for local use and re-entry into export markets	No realised or perceived negative impacts

Table 16. (Continued)

Guideline/strategy	In favour of market development	Against market development
(continued)	• Provides SA government plan for trade representation across the region in Indonesia, Malaysia and Singapore in the waste industry for SA resource recovery companies	
Recycled Fill Materials for Transport Infrastructure Environmental Instruction 21.6-2015 - *DIT*	• Developed to manage oils, asphalt planings, road base, concrete, ballast and timber sleeper generated and re-used on transport infrastructure maintenance and construction projects • Recycled aggregates permitted for pavement construction consist of inert products such as crushed concrete, brick, tile and other ceramics	The permitted application of concrete waste has to be less than 20% by mass
Guide for the Re-use or Disposal of Surplus Soil 2017 - *DIT*	• Provides a guide to DPTI staff on the appropriate re-use or disposal of surplus soil and other C&D waste materials, i.e., concrete • Specifies approval conditions for reusing soil as waste fill	No realised or perceived negative impacts
South Australia's Waste and Resource Recovery Infrastructure Plan - *GISA*	• Highlights the need for PwRC market analysis and development as well as data collection and management, and online web platforms to support new and expanded infrastructure projects • Suggests the following actions: a) Expanding existing recycling facilities to process growing volumes of C&D waste (e.g., concrete crushing), as well as recovering more waste going to landfill such as waste soil b) Consideration should be given to establishing a 'soil bank' and to developing consistent and better-coordinated systems and practices to enable cost-effective re-use of low-risk waste soils	No realised or perceived negative impacts

Guideline/strategy	In favour of market development	Against market development
(continued)	c) (e.g., clean fill and intermediate-level waste soil) for development, redevelopment and infrastructure projects d) Creating suitable transfer routes to waste and resource recovery facilities and ports to export markets e) Investing in equipment and facilities for waste compaction and bulk hauling to reduce the cost of transporting waste to end-markets f) Developing of markets is needed for new and expanded volumes of recovered recyclables g) Creating local demand for these materials through government procurement policies and CE processes h) Establishing accredited testing for product standards and performance to provide purchasers with confidence in the quality of the remanufactured products	

2.4.6.3. Standards, Codes and Specifications

The *SA EPA* has defined approval conditions for using C&D waste materials as waste-derived fill (WDF). The *DIT* provides specifications for using C&D waste PwRC in the Department's projects. Specification: Part R15 Supply of Pavement Materials outlines the requirements for the supply and delivery of recycled pavement materials to be used in construction projects. It also suggests that, where fresh materials have been specified in contract-specific requirements, the contractor may submit a proposal to use PwRC. Table 17 presents a summary of standards, codes and specifications related to C&D waste market development in SA.

Table 17. SA government document statements in favour/against PwRC market development

Standard/code/specification	In favour of market development	Against market development
The standard for the production and use of Waste Derived Fill 2013 - *SA EPA*	• Describes the processes and approval requirements of using three waste sources for potential use as WDF. These sources are the soil for direct re-use, C&D waste PwRC and a homogenous mineral-based industrial residue • WDF approval is subject to the existence of an immediate market and appropriate materials flow and stockpile management	No realised or perceived negative impacts
Specification: Part R15 Supply of Pavement Materials - *DIT*	• Specifies the requirements for the supply and delivery of materials including recycled pavement materials to be used in the construction of roadworks, bridgeworks, railways and other applications associated with construction • Indicates that, where fresh materials have been specified in contract-specific requirements, the contractor may submit a proposal to use PwRC	No realised or perceived negative impacts

2.4.6.4. Relevant Organisations

GISA, the *DIT* and the *SA EPA* are the government agencies that contribute to market development for waste materials in SA. The *GISA* is statutorily commissioned to develop a waste strategy for the state [32]. The agency's responsibility includes public and industry awareness of and participation in resource recovery and green industry; identification of targets for resource recovery development and identification of any obstacles or potential obstacles to the implementation of the targets or any significant risks associated with the implementation of targets; and establishment of criteria and methods for assessing the adequacy of the strategy and its implementation. Furthermore, the Green Industries SA Act 2004 defined the agency's functions including market development, identifying and improving business efficiencies and opportunities, and raising public and industry awareness of waste management, resource recovery and innovation, and the best practice methods thereof. It is also tasked to promote research and development about resource recovery and waste management and to establish a collaborative partnership with government and non-government organisations such as the research sector to promote waste management and resource recovery. Additionally, the *DIT* has published several guidelines and specifications for using C&D waste in the Department's projects.

2.4.7. Tasmania (Tas)

Tas currently has the fewest waste regulations and policies concerning end-market development in Australia. The main legislative documents that affect C&D waste market development in Tas comprise only one act and one regulation, and little support is provided through these two pieces of legislation. Currently, there are no standards, (updated) strategies, policies or guidelines that affect C&D waste creation and stimulation. In Tas, waste is always considered waste regardless of its value. The *Tasmania Environment Protection Authority (Tas EPA)* and the *Department of Primary Industries, Parks, Water and Environment* are the government agencies that are responsible for waste management in Tas.

2.4.7.1. Primary Legislation

The only two relevant pieces of legislation in Tasmania provide little support for market development. In the Environmental Management and Pollution Control Act 1994, re-using and recycling is encouraged; the subordinate regulation (2020) as part of the application for environmental approval states that alternative options for reusing and recycling is mandated. The main focus of the primary legislation is on controlled waste materials such as poisons, fish waste and sewage. A significant barrier to C&D waste market development is the lack of a state-wide landfill levy to incentivise recycling and re-using activities. Table 18 provides a summary of regulatory statements enshrined in Tasmanian legislation in favour of and against market development.

Table 18. Tas regulatory statements in favour/against PwRC market development

Act and regulations	In favour of market development	Against market development
Environmental Management and Pollution Control Act 1994 - *Tas EPA*	Waste re-using and recycling is among the objectives of this Act	Waste is always waste regardless of its value and its future fate
Environmental Management and Pollution Control (Waste Management) Regulations 2020 - *Tas EPA*	Environmental approval should include any alternative options for re-use, reprocessing or recycling of the waste	Waste is always waste regardless of its value and its future fate

2.4.7.2. Relevant Organisations

The *Tas EPA* and the *Department of Primary Industries, Parks, Water and Environment* are the government agencies that are responsible for waste management in this state.

2.4.8. Victoria (Vic)

Vic is a leading state in market development for C&D waste materials. *Sustainability Victoria (SV)* recently introduced the Buy Recycled directory, in which state PwRC suppliers are identified and featured. Despite the lack of a current waste strategy in Vic, the state agencies have published various guidelines and strategies that inform waste management and resource recovery activities and decisions across the state. The *Department of Environment, Land, Water and Planning (DELWP)*, Waste and Resource Recovery Groups, the *Victorian Environmental Protection Authority (Vic EPA)*, *SV* and *VicRoads* are the leading government agencies that play a role in developing C&D waste end-markets in Vic.

2.4.8.1. Primary Legislation

The Environment Protection Act 1970 relies on several sets of principles to effectively manage waste and improve market conditions for PwRC. These include wastes hierarchy, shared responsibility, product stewardship, and the improved valuation, pricing and incentive mechanisms principle. The Act also provides the regulatory context to establish the regional Waste and Resource Recovery Groups (WRRGs), which serve several objectives and functions. WRRGs are Victorian State Government statutory authorities with a legislative responsibility to develop their Regional Waste Resource Recovery Plans (RWRRIPs) to specify how to satisfy the waste and resource recovery needs of their region over 10 years; currently, there are seven WRRGs across Vic. The Act indicates the requirements for landfill levy rebates when waste received at a landfill is demonstrated to be recycled, reprocessed and recovered. The Sustainability Victoria Act 2005 sets out *SV* objectives and functions to promote sustainability in the use of resources with some focus on PwRC market development and stimulation. Table 19 provides a summary of regulatory statements enshrined in Victorian legislation in favour of and against market development.

2.4.8.2. Guidelines, Policies and Strategies

In the absence of a current waste strategy (the previous version became outdated in 2010) *SV, Vic EPA* and *DEWLP* published several guidelines and strategies that have a particular focus on PwRC end-markets development. In 2018, *SV* released the Statewide Waste and Resource Recovery Infrastructure Plan, which is an essential guideline for end-markets creation and stimulation over the next 30 years in Vic. The Plan contains several goals, strategic directions and recommendations for enabling PwRC utilisation. The strategic directions provide the pathway to further resource recovery across Vic. This agency also published the Victorian Market Development Strategy for Recovered Resources, which outlines how the state government and other stakeholders should contribute to creating and stimulation of end-markets. The EPA released a supporting guideline, Management and Storage of Combustible Recyclable and Waste Materials. Table 20 presents a summary of guidelines, policies and strategies related to C&D waste market development in Vic.

Table 19. Vic regulatory statements in favour/against PwRC market development

Act and regulations	In favour of market development	Against market development
Environment Protection Act 1970 - Administered by *Vic EPA*	• Adopts wastes hierarchy, shared responsibility, product stewardship, and improved valuation, pricing and incentive mechanisms principles: Pricing and incentive mechanisms principles should establish incentive structures, including market mechanisms, which enable persons best placed to maximise benefits or minimise costs to develop solutions and responses to environmental problems • Outlines the objectives and functions of Waste and Resource Recovery Groups including: a) Undertake waste and resource recovery infrastructure planning b) Facilitate efficient procurement of waste and resource recovery infrastructure and services for its waste and resource recovery region c) Integrate regional and local knowledge into state-wide waste and resource recovery market development strategies d) Educate businesses and communities to reduce waste going to landfill by using waste and resource recovery infrastructure and services efficiently e) Ensure its plans and programs are informed by local government, business and community and inform state-wide plans and programs • Sets out the industry waste reduction agreement specifications including the obligation to provide an economic assessment of the market for the wastes that can be recovered • Specifies Vic's waste management policy development criteria • Specifies how the landfill levy rebate may be claimed when waste received at a landfill is recycled, reprocessed and recovered. This rebate is only claimable in annual landfill levy statements	Considers resources to be recovered as waste

Act and regulations	In favour of market development	Against market development
Environment Protection Act 2017 - Administered by *Vic EPA*	• Presents the Vic EPA's new objectives and responsibilities and a new governance structure	No responsibility regarding market development is allocated to Vic EPA as the main government agency with environmental protection focus
Environment Protection (Resource Efficiency) Act 2002	• Aims to improve the operation of the provisions of the Environment Protection Act 1970 in dealing with waste management and resource recovery • Presents the functions of *SV*	No statement on waste end-markets development and stimulation
Sustainability Victoria Act 2005	• Presents *SV*'s functions including those directly related to waste end-markets development: a) Foster a stewardship ethos concerning the use of resources b) Promote throughout Vic waste avoidance, waste reduction and recovery, re-use, recycling of resources and best practices in waste management c) Encourage and promote the development and use of environmentally sustainable practices, markets, technologies d) and industries, including resource efficiency, energy efficiency, renewable energy and water e) Facilitate the uptake of fledgling technologies, industries, markets and practices in environmental sustainability, including demonstration projects f) Facilitate the development of voluntary environmental sustainability initiatives g) Develop and implement strategies to foster sustainable markets for recovered resources and PwRC h) Develop tools to measure, monitor and report on Government waste, water and energy targets	No realised or perceived negative impacts

Table 20. Vic guidelines and strategies in favour/against PwRC market development

Guideline/strategy	In favour of market development	Against market development
Statewide Waste and Resource Recovery Infrastructure Plan 2018 - *SV*	• Presents the results of economic and transport assessments with some recommendations as to reducing transport costs and increasing the resource recovery economic benefits • Identifies the main stakeholders and their role in the waste and resource recovery system including community, business, industry and government, manufacturers, waste and resource recovery industry, local governments, *SV*, *EPA*, *DEWLP*, and WRRGs, the *Victorian Planning Authority* and the federal government • Presents the Plan's strategic directions and goals (in 30 years) include the following: a) Goal 1: Landfills will only be for receiving and treating waste streams from which all materials that can be viably recovered have been extracted b) Goal 2: Materials are made available to the resource recovery market through aggregation and consolidation of volumes to create viability in recovering valuable resources from waste c) Strategic Direction 1: Prioritise viable recovery d) Strategic Direction 2: Reduce landfill reliance e) Strategic Direction 3: Aggregate materials f) Strategic Direction 4: Utilise land g) Strategic Direction 4: Evidence-based decision-making • Identifies opportunities identified in regional implementation plans such as facilitating joint procurements for waste and resource recovery infrastructure	No realised or perceived negative impacts

Guideline/strategy	In favour of market development	Against market development
Victorian Market Development Strategy for Recovered Resources 2016 - *SV*	• Supports industry to develop and test new products made from recovered materials and promote their uptake and to strengthen markets for existing products • The Strategy's strategic directions include: a) Improve the quality of recovered resources to support manufacturing b) Improve consolidation and aggregation of recovered materials to contribute to growth in manufacturing c) Improve the performance of products incorporating recovered resources d) Increase the use of products incorporating recovered resources e) Cross government coordination within an integrated, statewide waste management framework f) Adopt appropriate, evidence-based approaches to government intervention g) Capitalise on policy and market signals supporting resource recovery h) Proposes prioritisation methodology for market development of recovered resources	No realised or perceived negative impacts
Recycling Victoria: A New Economy 2020 - *DELWP*	Specifies Vic's CE goals as follows: a) Goal 1 - Design to last, repair and recycle b) Goal 2 - Use products to create more value c) Goal 3 - Recycle more resources d) Goal 4 - Reduce harm from waste and pollution	No realised or perceived negative impacts
A Circular Economy for Victoria - *DELWP*	Contains the following sections relevant to recovered materials end-markets: a) Improving materials productivity b) Getting more value from recovered materials c) Building a resilient waste and resource recovery system	No realised or perceived negative impacts

Table 20. (Continued)

Guideline/strategy	In favour of market development	Against market development
Guideline: Energy from Waste 2017 - *Vic EPA*	Outlines how the Environment Protection Act 1970 and associated statutory policies and regulations are applied to the assessment of proposals that recover energy from waste Reviews the assessment criteria for energy from waste activities	No realised or perceived negative impacts
Victorian Waste Education Strategy 2016 - *SV*	Some focus areas with relevance to PwRC end-markets in this strategy include: a) Increase the Victorian community and business perception of waste management as an essential service b) Increase community awareness of waste and support and encourage waste avoidance c) Improve resource recovery and reduce contamination d) Strengthen Vic's waste and resource recovery education capabilities	No realised or perceived negative impacts

Table 21. Vic government document statements in favour/against PwRC market development

Standard/code/specification	In favour of market development	Against market development
Registration of Crushed Rock Mixes: Code of Practice - *VicRoads*	Specifies the requirements for using crushed rock mixes in project infrastructures	No realised or perceived negative impacts
Selection and Design of Pavements and Surfacings: Code of Practice RC.500.22 - *VicRoads*	Specifies the requirements for using waste materials in building pavements	No realised or perceived negative impacts
Section 812-Production of Crushed Rock for Pavement Base and Subbase - *VicRoads*	Specifies the requirements for using crushed rock in pavement subbase and light-duty base	No realised or perceived negative impacts
Section 820-Crushed Concrete for Pavement Subbase and Light Duty Base - *VicRoads*	Specifies the requirements for using crushed concrete in pavement subbase and light-duty base	No realised or perceived negative impacts
Section 821- Cementitious Treated Crushed Concrete for Pavement Subbase - *VicRoads*	Specifies the requirements for using crushed concrete in the pavement sub base	No realised or perceived negative impacts
Section 407-Hot Mix Asphalt - *VicRoads*	Specifies the requirements for using waste materials in project infrastructures	No realised or perceived negative impacts

2.4.8.3. Standards, Codes and Specifications

VicRoads is a public agency that specifies the physical and chemical properties of several waste-derived materials in applications in infrastructure projects. Table 21 presents a summary of standards, codes and specifications related to C&D waste market development in Vic.

2.4.8.4. Relevant Organisations

The *DELWP*, WRRGs, the *Vic EPA*, *SV* and *VicRoads* are the main government agencies that play a role in developing C&D waste end-markets in Vic. The first four entities provide the *Victorian Government* with policy advice, environmental regulation, programs to implement policies and reporting on the state's environment. The objective of *SV*, as enshrined in the Sustainability Victoria Act 2005, is to facilitate and promote environmental sustainability in the use of resources. The agency has provided several guidelines and strategies to increase resource recovery activities across Vic. Meanwhile, *VicRoads* develops and administers specifications of PwRC to be used in the state's infrastructure projects.

2.4.9. Western Australia (WA)

WA is vast and located a considerable distance from waste end-markets, which can impact this state's investment in waste and recycling infrastructure and overall recycling rates. Despite this, WA has a heavily regulated waste management regulatory framework. The state regulatory framework considers waste as a resource and provides an exemption for the application of waste in construction and maintenance work in licenced landfills. The framework also provides regulatory supports for the application of a range of waste management strategies such as polluters pay, product stewardship, extended producer responsibility and end-users liability for material whole lifecycle costs. The organisations with the most impact on C&D waste market development in WA are the *Department of Water and Environmental Regulation (DWER), the Waste Authority*, the *WA Environment Protection Authority (WA EPA)* and *MainRoads WA*.

2.4.9.1. Primary Legislation

The Environmental Protection Act 1986 is the overarching piece of legislation that prevails over other state legislation. This Act distinguishes waste used for reclaiming purposes from waste. It specifies two waste strategies—the polluter pays principle and end-users responsibility—that aid market development for PwRC. The Act instructs allocation of state funds to re-using and recycling activities and favours market incentives. The second key act, the Waste Avoidance and Resource Recovery Act 2007, administered by *DWER*, specifies product stewardship and extended producer responsibility to encourage recycling activities. Table 22 provides a summary of regulatory statements enshrined in WA legislation in favour of and against market development.

2.4.9.2. Guidelines, Policies and Strategies

The *Waste Authority* released WA's waste strategy document, Waste Avoidance and Resource Recovery Strategy 2030, in which several actions and strategies are recommended to develop C&D waste PwRC. The agency has also published several other guidelines and statement positions that cover the PwRC market development. Table 23 presents a summary of guidelines, policies and strategies related to C&D waste market development in WA.

2.4.9.3. Standards, Codes and Specifications

Three agencies provide specifications for using PwRC in WA construction projects: the Waste Authority, *MainRoads WA* and the *Institute of Public Works Engineering Australia (IPWEA)*. In 2020, the Waste Authority published a state-wide specification for recycled road base and drainage rock. *MainRoads WA* also have a specification for C&D waste PwRC for application in pavement projects. Furthermore, *IPWEA*, in partnership with the *Western Australia Local Government Association (WALGA)*, set requirements for supplying C&D waste materials in road works. Table 24 presents a summary of standards, codes and specifications related to C&D waste market development in WA.

2.4.9.4. Relevant Organisations

The state organisations involved in C&D waste market development in WA are the *DWER*, the *Waste Authority*, the *WA EPA* and *MainRoads WA*. In WA, Acts and regulations are administered by DWER. The Waste Authority is responsible for developing and updating the state waste strategy document in which several actions and strategies are recommended to develop C&D waste PwRC. The agency also provides statement positions about various aspects of waste management in WA. The *Waste Authority*, *MainRoads WA* and *IPWEA/WALGA* have separately published specifications for using C&D waste-derived products in infrastructure projects.

Table 22. WA regulatory statements in favour/against PwRC market development

Act and regulations	In favour of market development	Against market development
Environmental Protection Act 1986 - *DWER*	• Does not consider all waste materials as waste • Enforces the polluter pays principle: Those who generate pollution and waste should bear the cost of containment, avoidance or abatement • The users of goods and services should pay prices based on the full lifecycle costs of providing goods and services • Indicates that environmental goals should be pursued most cost-effectively, by establishing incentive structures including market mechanisms, which enable those best placed to maximise benefits and/or minimise costs to respond to environmental problems • Indicates that amounts of money collected from waste penalties and landfill levy may be applied to fund programmes relating to the management, reduction, re-use, recycling, monitoring or measurement of waste that are approved by the Minister	No realised or perceived negative impacts
Environmental Protection (Landfill) Levy Act 1998 - *DWER*	• No realised positive impacts	No realised or perceived negative impacts
Waste Avoidance and Resource Recovery Act 2007 - *DWER*	• Adopts the waste management hierarchy • Promotes the most efficient use of resources, including resource recovery and waste avoidance • Specifies product stewardship plans and extended producer responsibility schemes • Establishes Waste Authority functions such as promoting market development for recovered resources and PwRC	No realised or perceived negative impacts No specific information about C&D waste stream

Table 22. (Continued)

Act and regulations	In favour of market development	Against market development
Waste Avoidance and Resource Recovery Regulations 2008 - *DWER*	• Describes the exemption conditions for C&D waste as follows: a) For the waste used for construction and maintenance work carried out on the licensed landfill b) Uncontaminated soil or other clean fill received at a licensed landfill for land reclaiming purposes on landfill site c) Waste that is not disposed of to landfill but is collected and stored at a licensed landfill for recycling	No realised or perceived negative impacts
Environmental Protection Regulations 1987 - *DWER*	• Similar to the overarching Act	No realised or perceived negative impacts
Landfill Waste Classification and Waste Definitions 1996	• Defines waste classes, C&D waste sub-streams and clean fill	No realised or perceived negative impacts

Table 23. WA guidelines and strategies in favour/against PwRC market development

Guideline/strategy	In favour of market development	Against market development
Waste Avoidance and Resource Recovery Strategy 2030 - *Waste Authority*	• Recommends the following strategies: • Implement sustainable government procurement practices that encourage greater use of PwRC and support local market development • Prioritise local market solutions for recyclable materials • Identify and implement options for collaboration between industry and the State Government to support market development and recovery with an emphasis on focus material (e.g., C&D waste) • Provide funding to promote the use of priority PwRC and support the establishment of local markets with an emphasis on focus materials • Support community, government and industry initiatives that promote resource recovery • Develop a legislative framework to encourage the use of waste-derived materials, including product specifications, to build confidence in PwRC, increase their demand and develop relevant markets while protecting the environment	No realised or perceived negative impacts
Waste Data Strategy 2019 - *Waste Authority*	• Undertake annual data collection activities to monitor waste and recycling performance • Collaborate with other jurisdictions and agencies to develop standards for collecting more detailed data about waste flows in the CE • Develop policies and legislation in consultation with stakeholders to increase the accuracy of waste data	No realised or perceived negative impacts
Waste Authority Position Statement Construction and Demolition (C&D) Waste Position Statement 2016 - *Waste Authority*	• Communicates the benefits of C&D waste recycling • Supports further research to identify suitable applications for C&D derived products	No realised or perceived negative impacts
Source Separation of Waste Position Statement 2014 - *Waste Authority*	• Communicates the benefits of C&D waste recycling • Encourages source separation for C&D waste materials	No realised or perceived negative impacts

Table 24. WA government document statements in favour/against PwRC market development

Standard/code/specification	In favour of market development	Against market development
Roads to Re-use: Product Specification - recycled road base and recycled drainage rock- 2020 - *Waste Authority*	Sets out product specification for recycled road base and recycled drainage rock Ensures that material meeting this specification will no longer be considered waste for the waste levy	No realised or perceived negative impacts
IPWEA/WALGA Specification for the supply of recycled road base 2016	Specifies C&D waste PwRC properties to be used in road base	No realised or perceived negative impacts
MainRoads Western Australia, Specification 501: Pavements 2020	Specifies supply of C&D waste PwRC properties to be used in pavement layers	No realised or perceived negative impacts

Conclusion

This review has demonstrated the differences and similarities between the Australian state and territory regulatory frameworks concerning the development and stimulation of end-markets for C&D waste recyclables. This study analysed 102 federal, state and territory government policies, standards, guidelines, legislation and specifications to determine the extent to which the Australian regulatory frameworks are in favour of or against C&D waste market development. The following are the key findings of this analysis:

1) Most of the primary state and territory legislation does not define C&D waste and considers waste as always waste regardless of its value or future usage;
2) Specifications for application of C&D waste PwRC vary between states and territories, which has a negative impact on interstate market development and stimulation;
3) In most waste strategy documents, the significance of waste market development is highlighted, and some waste strategy documents suggest recommendations to improve it; and
4) Among the states and territories, NSW and Tas have the most extensive and smallest regulatory framework relevant to C&D waste management and market development thereof, respectively.

Chapter 3

The Capacity of Existing End-Markets in Australian Jurisdictions

3.1. Introduction

C&D waste is the largest source of waste in Australia. The latest records [2]. show that the annual quantity of C&D waste produced in the eight states and territories of Australia is about 27 mt, which is followed by commercial and industrial (C&I: 21.9 mt) and municipal solid waste (MSW: 12.6 mt). Management of such a quantity within Australia requires the adoption of a CE approach that is underpinned by resource efficiency and waste recovery enhancement. Market development has been identified as a sustainable option that ideally engages all stakeholders who have a role to play in the construction materials supply chain and extension of their lifecycle beyond their end-of-life [22].

This chapter aims to investigate the factors that impact development and the operation of the existing markets for C&D waste PwRC. The objective of this study is to assess the capacity of existing end-markets in Australian jurisdictions, how the existing markets can be stimulated and how new end-markets can be created for sustainable management of C&D waste.

3.2. Approach

The operation of end-markets for C&D waste materials is subject to multiple factors. This chapter, therefore, focuses on the major factors identified as having an impact on the creation and stimulation end-markets for the C&D waste stream. As waste regulation occurs at the state and territory level, the research methodology is designed to individually analyse market conditions across the Australian jurisdictions. To define the study scope, five case materials were selected for detailed market analysis: concrete, brick, steel, glass and asphalt. The materials were selected by the project industry and government partners from a list of different construction materials using five criteria: the embodied energy, embodied water, embodied greenhouse gas emission, the rate of annual waste and whether the material is considered a priority resource in each state and territory.

3.2.1. Data Collection and Processing

A desktop literature review was employed to capture information regarding the operation of existing markets and the development of new markets for C&D waste PwRC in Australian states and territories. The document analysis involved reviewing government and industry reports, research publications, industry magazine articles and articles published through online news agencies. To reflect the status quo of market development in Australia, only recent information that was released after 2016 was considered. In total, 121 unique references were used to shape this chapter.

Waste data were primarily extracted from the latest national waste dataset that was released in late 2020 [2]. This dataset includes aggregated and individual waste data both nationally and jurisdictionally and indicates the waste fates (e.g., disposal, recycling and energy from waste recovery). However, in this dataset, the quantity of each waste material generated in states and territories is not specified. Hence, the authors assumed that the total waste generated is equal to the sum of the quantity of waste recycled and landfilled. Furthermore, the recommendations provided for the stimulation of the existing end-markets in each state and territory were primarily based on the experts' views published in different reports such as *Environment and Communications References Committee*'s report [24]. and the issues identified for market development in the states and territories.

3.3. Analysis of Existing C&D Waste End-Markets: Australia

3.3.1. Introduction

This section presents an analysis of the capacity of existing C&D waste end-markets in Australia. The analyses are based on C&D waste statistics, the economics of C&D waste and existing markets for C&D waste PwRC.

3.3.2. C&D Waste Data

The C&D waste streams are the largest source of waste in Australia. The quantity of waste materials generated in C&D waste streams has increased since 2016 [2]. The latest data shows that in the fiscal year of 2018–19 about 22.7 mt of C&D waste was generated (Figure 2), which was the largest contributor to the total solid waste (91.7%) in Australia.

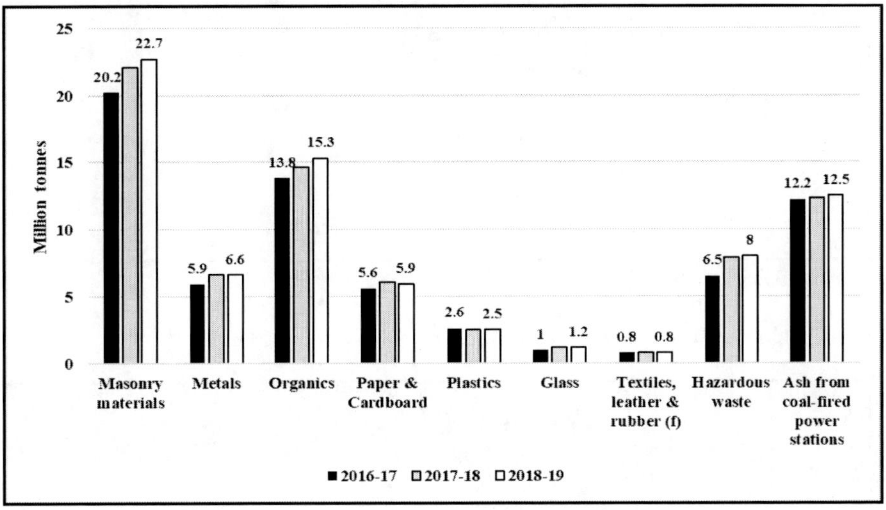

Source: ABS (2020) [33].

Figure 2. Waste generation by waste material in Australia (2016–2019).

In 2018–19, the construction and manufacturing industries generated the largest quantity of waste (12.8 mt) followed by electricity, gas and water & waste services (12.3 mt). The waste quantity in the construction sector has increased by 23% since 2016-2017 (Figure 3).

In terms of waste recovery activities, C&D waste materials ranked first among other waste streams (Figure 4). In the fiscal year of 2018–19, more than 81% of C&D waste generated was recovered throughout Australia, followed by metals, and the paper and cardboard category (66.9%). During the same period, Australia exported 6% of the total waste generated in the country.

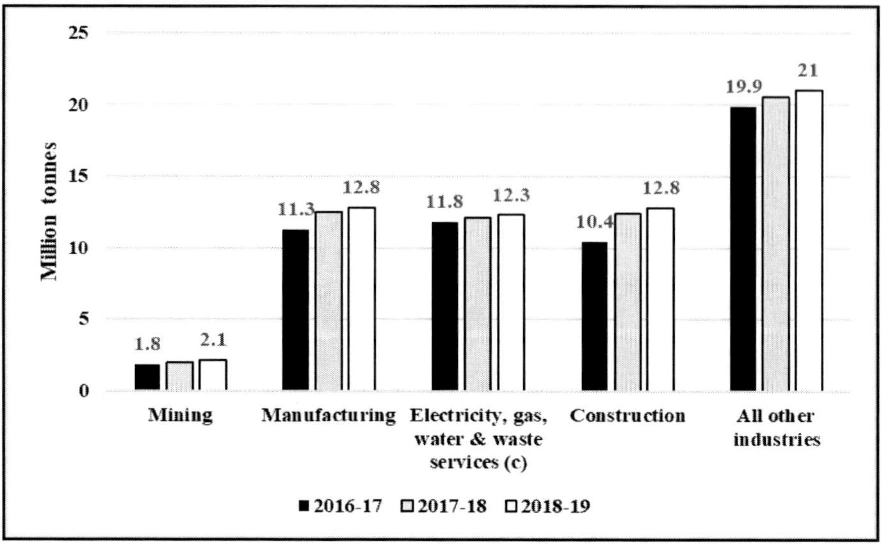

Source: ABS (2020) [33].

Figure 3. The waste generation by industry in Australia (2016–2019).

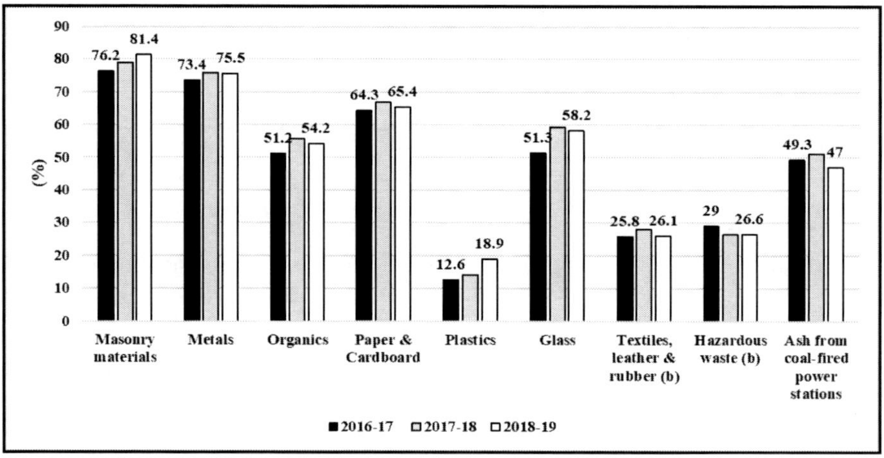

Source: ABS (2020) [33].

Figure 4. The waste recovery rate for various waste streams.

3.3.3. The Economy of C&D Waste

The waste management and resource recovery sector contribute 0.3% of Australia's GDP. In 2019, 36,000 jobs were recorded in this sector, an increase of 16% from 2016. The sector added $4,866 m to the economy. The ABS reported that, among various sectors, the construction industry spent $2 bn on waste collection, treatment and disposal services [33]. These expenditures were calculated through the indicator of waste intensity, which quantifies the amount of waste generated per million dollars of value-added to the economy (Figure 5). For the construction industry, this indicator is 87 t/m dollars. These figures reveal the significant magnitude of the C&D waste economy and the potential for the development of domestic markets.

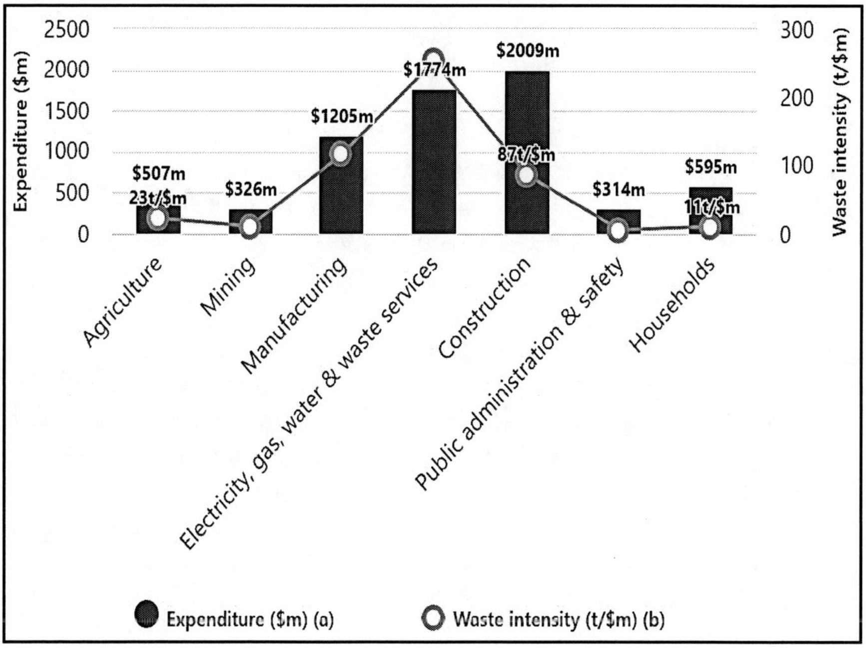

Source: ABS (2020) [33].

Figure 5. Expenditure on waste services.

3.3.4. The Market for C&D Waste PwRC

Market development activities for C&D waste vary significantly across the Australian states and territories. However, most Australian jurisdictions have now acknowledged the value of a CE in the construction industry and sought to transition to non-linear construction material end-of-life management. In some states, such as Vic, WA and NSW, large public infrastructure projects have unlocked the potential of using C&D waste PwRC in construction projects and present an opportunity to develop sustainable end-markets for these materials. The following review of market conditions in each state and territory will provide an insight into the main factors affecting local market creation and stimulation. It can be used to establish a national platform to facilitate C&D waste trading across Australia.

3.4. Analysis of Existing C&D Waste End-Markets: Australian States and Territories

3.4.1. Operation of C&D Waste End-Markets in the ACT

The ACT has a population of 400,000 residents living in about 145,000 households and is one of the fastest-growing regions in Australia. This territory is one of the leading jurisdictions in waste management, with about 70% of waste being reused and recycled. In 1996, the ACT was the first jurisdiction in the world to employ a 'No Waste' strategy; this important initiative promoted the idea of consideration of waste as a resource between 1996 and 2000. Governance and regulation of C&D waste in the ACT is performed by the ACT EPA and Transport Canberra and the City Services Directorate. In 2018, the ACT government set a 95% target to recover C&D waste in this territory. The rate of waste generation has a linear relationship with construction activities. Since 2014, the value of the construction industry in the ACT has surged, and the total value of work done reached AUD $766, 681 in 2020 [33]. (Figure 6), approximating a more than 36% increase from the corresponding 2014 value.

This section presents data relevant to the waste management of five priority construction materials, an overview of the existing markets, the conditions of material recovery facilities (MRFs), some case studies of using PwRC in construction projects, the extent of government support and recommendations to enhance market operation across the territory.

3.4.1.1. C&D Waste Data

The latest report commissioned by the *DAWE* [2]. showed a growth in the overall rate of ACT waste generation, which was less than that in population growth. The territory government had set a target of 90% resource recovery by 2025 [34], which it later increased to 95%. The statistics regarding priority C&D waste materials in the ACT are shown in Table 25.

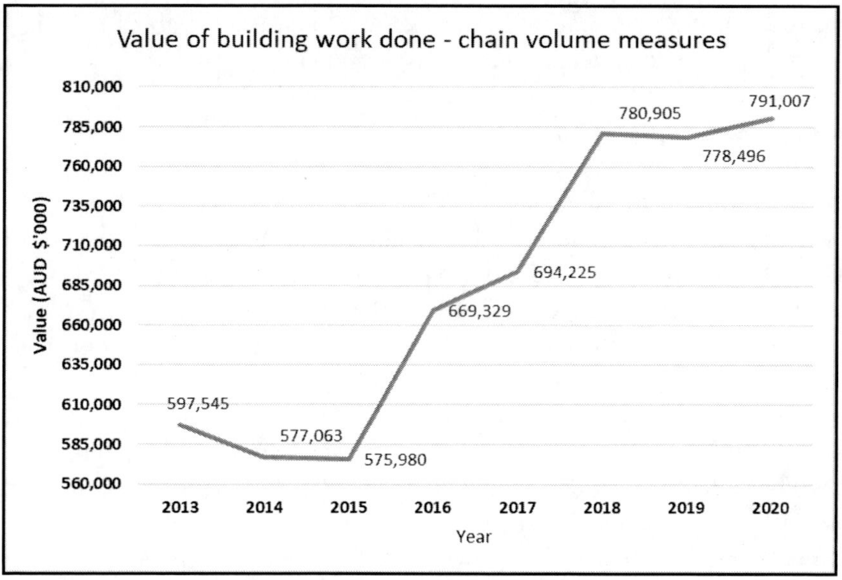

Source: ABS (2020) [34].

Figure 6. The value of construction activities in the ACT 2013–2020 (until Sep).

Table 25. Priority C&D waste materials statistics in the ACT

Material	Waste generation (t)	Disposal (t)	Recycling (t)
Brick[a]	0	0	25,098
Concrete	116,646	1,177	115,469
Asphalt	12,009	0	12,009
Steel *	0	0	0
Glass *	8	8	0

Note: a recycling figure is for the total waste stream; * numbers are for C&D waste streams only.
Source: 2020 National Waste Database [35].

It is worth noting that there is a significant waste exchange between ACT and NSW. For example, about 562 kt of Queanbeyan's (NSW) disposable C&D waste in 2017 was sent to the ACT for re-landfilling; at the same time, C&D waste that was originally generated in the ACT was shipped to NSW because it is a shorter distance than to ACT facilities and offers disposal in non-levy landfills [36]. In 2018, the Waste Management Review [37]. reported that the NSW EPA has partnered with police, the ACT Government and local councils to target rogue operators supplying waste soil from construction sites advertised as clean fill to unsuspecting property owners. Compliance and roadside checks were part of the crackdown to ensure fill going to a site had the appropriate council approval for its acceptance.

3.4.1.2. Existing Markets

The market for certain C&D waste materials such as concrete, brick, steel and glass are well-established in the ACT. Table 26 provides a summary of the conditions of the market for the case study C&D waste materials in the ACT. The information is sourced from the existing literature and secondary data available publicly. Conversely, the market for EfW (Energy from Waste) recovery (extraction and combustion of methane gas) has been diminished due to a change in government policy following public concerns about the pollution associated with the operation of EfW facilities, notably West Belconnen Resource Management Centres and Mugga Land Landfills [38]. Currently, the Business Recycling platform [39]. provides critical information on C&D waste recyclers in the Capital Region.

3.4.1.2.1. Material Recovery Facilities

A 2018 report [39]. on the ACT waste management system indicated that the recent increased waste landfill levy has resulted in a massive diversion of C&D waste materials (>80%) to Material Recovery Facilities (MRFs). Since 2004, the ACT has recovered 65–74% of this waste stream. The report also specified that on-site waste separation by construction companies could reduce the cost charged at MRFs from $130/t to $20/t.

The ACT government has ruled out EfW as a waste management option for the territory in favour of climate change direction objectives. In the ACT, from 2020, the new MRFs that intend to engage in EfW extraction activities have been prohibited from incineration, gasification, pyrolysis or variations of these for energy recovery [41]. This policy limits the market availability for combustible C&D waste such as timber.

The policy also requires:

- Evaluation of the operation of existing recycling facilities;
- Evaluation of issues impacting recycling facilities; and
- Identifying strategies to overcome these issues.

Table 26. The market for case study materials in the ACT

Material	Types/Application	Description	End-users
Brick	Brick dust, crushed brick and whole brick. The whole brick is cleaned and reused; crushed brick is used as a raw material for brick production, generate fill, drainage and road pavement.	The market for this material is quite strong in the ACT. Brick and tiles are the second-highest PwRC in the Capital Region.	The main clients are private builders, civil and infrastructure contractors, and landscaping suppliers
Concrete	Major concrete-based products include crushed concrete, crushed dust, aggregated and sub-base. Crushed concrete is used as a substitute for virgin aggregates (e.g., limestone, rock and gravel).	Concrete recycling involves the utilisation of crushing machinery that crushes concrete into various sizes. There is a strong market within the Capital Region for recycled concrete for road base and fill and for drainage lines as aggregate. Traditionally, the demand for recycled concrete in civil, commercial and residential activities outweighs its supply generated from construction activities.	The main end-users include private building contractors, civil and infrastructure contractors, and wholesale landscape suppliers
Asphalt	Asphalt millings	There is a well-established recycling process and market for asphalt in the ACT.	
Steel	Steel scrap, rebar and stainless steel	There is a mature market for ferrous metals, including steel, in the ACT. Builders, skip operators and demolition contractors separate steel from waste mix-loads. Hence, there is a high level of recycling of steel in the Capital Region. The approximate figure for buying steel waste by recyclers is $250; along with the cost-saving associated with avoiding landfill levy fees, this induces strong economic incentive for steel recycling, and hence the market is quite viable for this material. This waste source is the third most PwRC in the Capital Region.	Scrap metal merchants are the main clients for recycled steel

Source: Canberra Business Chamber (2014) [40].

Canberra Concrete Recyclers Pty Ltd [42]. is an active recycler that accepts a wide range of C&D waste in the Capital Region. This company structures its operation costs based on the mix of the load it receives at the weighbridge. Tables 27 and 28 outline the charges associated with C&D waste acceptance at a recycling facility based in the ACT and provide cost information for a range of PwRC. The figures in the tables show that tipping fees at recycling facilities are much cheaper than landfilling fees, which are currently $199.2 and $146.2 in metropolitan and regional areas, respectively.

Table 27. Tipping fees and PwRC price list example in the ACT- Part I

Waste material	Tip fee ($)/t	Waste material	Tip fee ($)/t
Brick/Terracotta tiles/pipes clean	23	Concrete oversize (+600- +800), including blocks, piers, poles, beams	16-100
Clean concrete (± 400 mm)	12-14	Coloured Tiles (excl. White tiles)	23
Brick and Concrete Mix	30	Asphalt	6
Virgin excavated natural material (VNEM)	41	Demolition waste, all categories	160
Products			
Asphalt subbase (20 mm)	20	Compactable fill	8
Asphalt oversize (40 mm)	18	Top soil (unspecified)	16
Road-base/sub-base -20mm DGS	16	Bricks unclean	35
Aggregate (5, 14, 20 mm)	16-29	Uncrushed screened concrete	27
Crushed concrete dust (7, 14, 20 mm)	12.5-16	Landscaping rocks	25

Source: Canberra Construction Recyclers (2020) [43].

Table 28. Tipping fees and PwRC price list example in the ACT- Part II

Waste material	Tip fee ($)/t	Waste material	Tip fee ($)/t
Drilling fluid	60.50	Rock	7.70
Non-clean fill	66	Clean concrete	9.90
		Contaminated concrete	110
VNEM	24.75	All metals	Nil
Asphalt	7.70	Clean brick	7.70
Mixed masonry-Terra/Ceramics/Coloured Brick (not Clean Red Brick)/Concrete/Natural Marble/Natural Granite)	22	C&D waste "A" Grade (primarily recyclable materials, i.e., timber, brick, asphalt)	132
		C&D waste "B" Grade (primarily non-recyclable materials, i.e., plastics, plywood, cardboard, fence palings)	154

Waste material	Tip fee ($)/t	Waste material	Tip fee ($)/t
Products			
Topsoil (screened)	15.40	Aggregates 10-60 mm	15.40
Topsoil (unscreened)	3.30		
Scalps (brick)	Free	Asphalt millings	17.60
Scalps (Dark)	11	Crushed brick: 5-7 mm 10-20 mm	 17.60 33
Concrete backing	15.40	Recycled golden mulch (m^3)	15.40
Brick dust/crusher dust	3.30	Rock	44

Source: ACT recycling (2020) [44].

3.4.1.2.2. Use of PwRC in Construction Projects

The following is an example of use of PwRC in construction projects in the ACT. The case study is representative of market operation in the Capital Region.

Case study 1: Road asphalt containing recycled materials [45] – In March 2019, the ACT laid around 1,000 t of asphalt containing PwRC during an ACT roads trial on sections of Horse Park Drive and Gundaroo Drive (Figure 7). The asphalt contains several PwRC including 250 kg of RAP. The *ACT Government* used this product in other road infrastructures. The cost of using this product is equivalent to that of traditional asphalt. The impact on local government budgets of not using PwRC could also contribute to higher waste disposal costs.

Source: Riqtact (2019) [46].

Figure 7. ACT roads trial on sections of Horse Park Drive and Gundaroo Drive.

3.4.1.2.3. Government

In 2020, an agreement between the federal government and the *ACT Government* was reached to deliver a major upgrade to MRF through financial support under the former's $190 m Recycling Modernisation Fund [46]. The federal government committed to providing $21 m to the ACT MRF. The objective of this fund is to better separate and process recycling streams. In this way, it will enhance market development and stimulation for waste materials, including C&D waste sources. Further, the *ACT government*—in pursuit of its 95% target for C&D waste recovery—plans to deliver exemplars of waste resource recovery in the construction and operation phases of Canberran projects [47].

3.4.1.2.4. Recommendations

A report prepared for Canberra Business Chamber makes some recommendations to improve the C&D waste management system in the ACT, some of which are relevant to the scope of this study. Furthermore, the National Waste Data and Reporting (cycle 2017–19) report [48]. also offers some suggestions to improve the C&D waste market. The following are the selected recommendations from these two reports:

- Establish a website or mobile application to develop a market for exchange in the soil (the most relevant to the scope of this study).
- Establish a regional website or mobile application to provide information about the location, opening hours and restrictions of all landfill sites.
- Require tenders for government contractors to have criteria that favour the recycling of materials.
- Determine and fix a standard and protocol for the collection of C&D waste data.
- Revisit the ACT's EfW policy to reinstate the government support for EfW activities for combustible C&D waste materials.
- Require responsible organisations to account for solar panels as an emerging issue.

3.4.2. Operation of Existing C&D Waste End-Markets in NSW

NSW is a south-eastern Australian state, distinguished by its coastal cities and national parks. It has a total area of 800,642 km² and covers 10.4% of Australia. NSW also has the largest population in Australia. Based on predictions, NSW's population will reach 8.34 m by the end of June 2021 [49]. The construction industry is required to provide residential builds and infrastructure to accommodate and provide urban amenities for the growing population. The construction industry, therefore, plays an important role in the state's economic profile. The industry is the fourth highest contributor to NSW's gross state product (GSP). Except for 2019, the value of building work done (all construction sectors and building types) in NSW (Figure 8) has steadily grown over the last few years [34], reaching $10.68 bn in 2018. Notably, NSW has commenced work on the largest infrastructure upgrade program in Australia. The state government has committed to spending $80 bn on new economic infrastructure over the four years from 2020 [50].

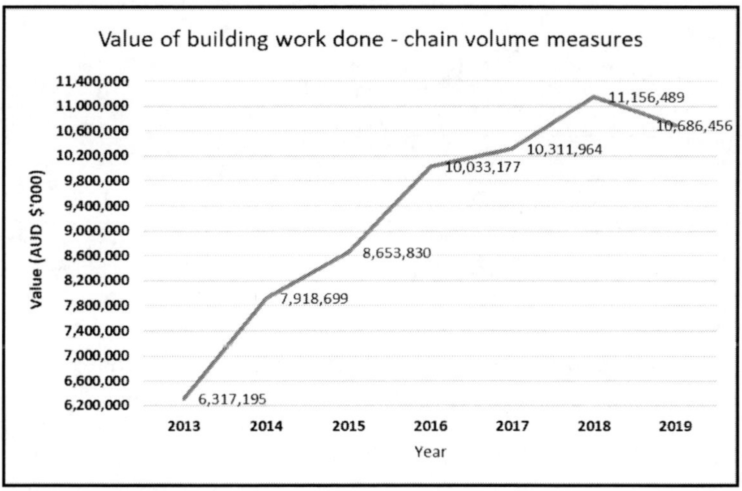

Source: ABS (2020) [34].

Figure 8. The value of construction activities in NSW 2013–2020 (until Sep).

3.4.2.1. C&D Waste Data

The latest report commissioned by the DAWE shows that the five priority materials in this report were recycled in significant quantities in NSW in 2018–19 (Table 29). Brick and concrete are the primary waste materials and were recycled in an almost equivalent amount. In a recent report on the state's waste sector, no waste disposal data is recorded for the fiscal year 2018–19. However, the collective trend in C&D waste generation shows an increase in the quantity; it is likely that NSW's significant infrastructure program and its spill over into private sector investment has contributed to this growth [51].

In NSW, about 7,752 kt of C&D waste was recycled in 2018–19, which is 3% down from 2016– [17]. Similarly, the C&D waste disposal quantities witnessed a 3% decrease compared to 2016–17, reaching 1,905 kt. In 2018, C&D waste was the largest source of waste, accounting for 60% of waste generation volume [52]. In contrast, EfW records show a significant increase in energy recovery between 2016–17 and 2018–19; the state recovered energy from 40 kt of C&D waste, which is a 288% increase compared to the previous data reporting period.49 Since 2007, the resource recovery rate in the state has followed a rising trend. The latest estimated waste recovery rates for NSW are 68% (recycling rate) and 77% (recycling + energy recovery rate) [2]. The government's recycling target rate for C&D waste streams in NSW is 80%.

Table 29. Priority C&D waste materials statistics in NSW

Material	Waste generation (t)	Disposal (t)	Recycling (t)
Brick[a]	1,291,546	0	1,291,546
Concrete	1,291,339	0	1,291,339
Asphalt	85,275	0	85,275
Steel *	0	0	547,471
Glass *	995	0	995

Note: a recycling figure is for the total waste stream; * numbers are for C&D waste streams only.
Source: 2020 National Waste Database [36].

Table 29 indicates that asphalt has found its way to market; indeed, the government's support—focused on using PwRC in road constructions—has unlocked the potential of application of reclaimed asphalt in NSW road infrastructure [52].

3.4.2.2. Existing Markets

NSW's focus on the development of the PwRC market tends to have prioritised municipal waste materials. However, in recent years, with changes in policies and the introduction of the CE concept, coupled with increased construction activities and elevation of its quantity [2], addressing C&D waste streams is gaining traction among policymakers. The government, in collaboration with the private sector, intends to increase resource efficiency in the construction industry through market development and stimulation.

Reflecting this shift is the NSW Government's (*Department of Planning, Industry and Environment*) partnership with Planet Ark to launch Australia's most comprehensive online database that helps small to medium businesses find recycling collection and support services for the workplace. This platform [40], which was established in 2019, invites providers of recycling and waste disposal services to list their details for free.

The market operation relies heavily on private businesses and recycling facilities that individually contribute to C&D waste trading; online versions of these have become popular recently. The volume of construction activities and the quantity of C&D waste generated and recycled in NSW has provided a strong motive for recycling businesses to create and stimulate markets for such products.

For instance, the Recycled Building Centre, which is a Sydney-based recycler, sells reclaimed building waste materials from the demolition of houses, factories and warehouses. The major share of their materials belongs to bricks, timber, sandstone and steel. The company has an online shopping platform to sell these products. Another recycler, the Building Recyclers Depot PTY Ltd [53], sorts and sells C&D waste based on their previous/future application and function, i.e., kitchen, bathroom, windows and doors, and landscaping. In addition to their webpages, these recyclers also use other popular trading platforms such as Gumtree and Facebook Marketplace to trade their products. Table 30 shows a (live at the time of writing) price list of PwRC that are supplied in the NSW metropolitan area.

Table 30. PwRC price list in NSW

PwRC	Turtle Nursery and landscape supplies Price (A$/t)*	PwRC	Gumtree Price (A$/t)*
Recycled Roadbase	46	Recycled Roadbase	35
Recycled concrete aggregate (10, 20 mm)	55	10 mm recycled brick Aggregate	44
Recycled asphalt	46	10 mm recycled brick aggregate (bulk bag)	55
Recycled concrete dust	45	Recycled concrete dust	10
Recycled concrete aggregate 40/70	55	Recycled concrete aggregate 40/70	20

Notes: * Prices are current as of January 2021.

Concrete—To estimate the potential market for recycled concrete products in NSW, the following calculations are made. Assuming all recycled concrete is to be converted to 10 mm (concrete), which is priced at $55/t (Table 30), and accounting for 10% residual waste during recycling, the total potential market for recycled concrete is $63.9 m/year.

Total potential market size = the state's concrete recycling quantity – 10% residual waste allowance × the retail price of recycled product

Total market worth = $1,291,339 – $129,133 × $55 = $63,921,330

Brick—Using the above equation and considering 10% residual waste for 10 mm brick aggregate and recycled asphalt, the total potential markets are $51.1 m and $3.5 m, respectively. The potential market for steel and glass is $690 m and $7.96 m, respectively, based on a retail price of 1.26/kg [54]. and 8/kg [55], respectively (Figure 9).

Figure 9. The potential market for priority materials in NSW.

3.4.2.2.1. Material Recovery Facilities

Despite the C&D waste recycling industry in NSW being considered to be mature and the state's overall strategy to support diverting waste from landfills, MRFs lag behind the state's ambitious waste recycling target, set at 80% [56]. A report by the NSW EPA shows that many licensed C&D recycling facilities are diverting from landfill well under 50% of the waste they receive, representing a significant loss of valuable resources from the productive economy [57].

A recent report by PWC [52]. indicates that the C&D waste recycling rate throughout the state in 2018 varied between metropolitan levy areas (81%), non-levied areas (34%) and regional levy areas (64%). The low recycling rate in non-levied areas highlights the need to impose a landfill levy that will eventually result in further recycling and market development.

3.4.2.2.2. Use of PwRC

NSW is a pioneer in the testing and application of PwRC in pavements and road constructions. Since 2008, the state has invested heavily in trials demonstrating the suitability and performance of PwRC in infrastructure projects [57]. Below are two case studies in which multiple organisations cooperated to encourage market development for MRF products.

Case Study 1: Crushed glass in pavement construction [58]. – In 2010, *Waverley Council*—in partnership with NSW *Department of Planning, Industry and Environment*; *NSW Roads and Traffic Authority*; *IPWEA*; and the Packaging Stewardship Forum—provided the first site within NSW to demonstrate alternate use of RCG in pavement construction as an accepted product in NSW roads (Figure 10). Waverley Council substituted 15 t of glass cullet into the road projects, 7.5 t into the asphalt and 7.5 t into concrete.

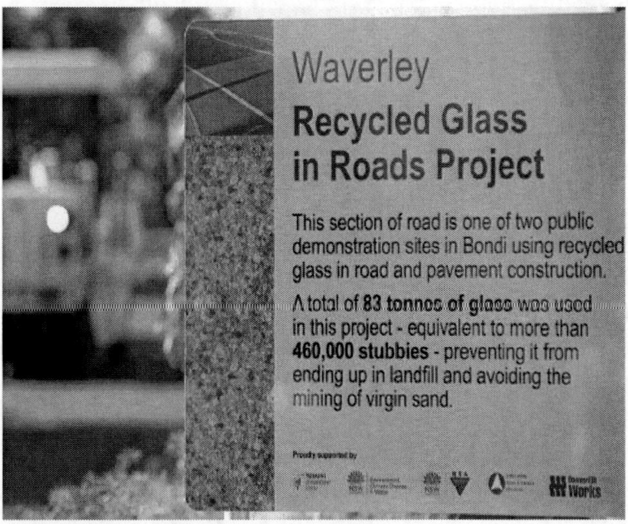

Source: Edge Environment Pty Ltd. (2012) [59].

Figure 10. Demonstration of recycled crushed glass in pavement construction.

Case Study 2: Asphalt construction for heavy vehicle load [53] – In 2018, Downer Pty Ltd, in collaboration with one council (i.e., *Randwick Council*) and other organisations, used asphalt in some NSW roads (selected roads in

Randwick and Maroubra) with heavy vehicle load traffic (Figure 11). The project objectives were to minimise reliance on virgin materials and reduce carbon emissions and the amount of waste being landfilled. From the C&D waste stream, RAP and coarse aggregate were used. The produced asphalt contained 20–30% RAP. The performance testing of the product shows that it improves fatigue, has superior deformation resistance for heavy vehicle loads, last 15% longer than standard asphalt and is capable of withstanding a 20% increase in traffic loading.

Source: Cheng (2020) [59].

Figure 11. Randwick Council trial for the use of recycled glass asphalt mix on three of its roads.

3.4.2.2.3. Government

The government sector in NSW has a pivotal role in creating and stimulating the market for using PwRC. The government, through various initiative and grants, has provided support for market development. As indicated earlier, the government support of the Business Recycling platform is a practical facilitating approach that directly impacts market development. Another initiative is the EPA's Circulate Industrial Ecology Program that supports the transition to a CE in NSW by funding innovative commercial industrial ecology projects that recover materials.

An initiative that is more specific to the C&D waste market is the Civil Construction Market Program, which provides grants to organisations to divert C&D waste from landfill and produce post-consumer recyclate such as glass, paper and plastic through reuse, recycling and industrial ecology projects in the NSW civil construction sector [60]. The grant objectives are to reduce the amount of C&D material sent to landfill, create new markets and opportunities for C&D waste material within NSW civil construction projects, create new markets and opportunities for post-consumer recyclate within NSW civil construction projects and remove the risk associated with trialling innovative resource recovery methods [61].

Furthermore, the government sector provides guidelines that aim to stimulate the market for C&D waste PwRC in construction projects. For instance, in 2020, *Local Government NSW*, in partnership with the *University of Sydney*, prepared and released a guideline [53] to drive the use of PwRC in roads and pavements in local councils, address the concerns preventing the use of PwRC by local councils and thus promote national uniformity and good practice in the specification and application of material reuse in roads and pavements by local council engineers. In addition to Local Government NSW, other organisations have supported case studies of using PwRC in construction projects; these include the *IPWEA (NSW)*, the *Roads & Transport Directorate, TfN*, *NSW EPA*, *Environment, Energy and Science (EES)* and the *NSW Environmental Trust.*

3.4.2.2.4. Recommendations

The following are some recommendations for the creation and stimulation of markets for C&D waste PwRC in NSW. These recommendations are extracted from experts' views published in various sources [53]:

- Deliver creative incentives such as fee waiver programs for local governments and businesses that use PwRC /process recyclables.
- Provide support via the waste levy for recycling, such as transport subsidies for regional areas, until market failures are addressed, rather than leaving everything to the market.
- Mandate minimum recycling content in all large-scale construction projects.
- Share recycling market information.
- Create markets for PwRC at the design and tender stages of state significant developments.
- Give priority access to government markets if companies can demonstrate above target resource recovery levels.
- Phase in *NSW Government* PwRC procurement targets and associated baselines, monitoring and reporting systems.
- Provide support to Local Government Procurement to enhance the Sustainable Choice Database to include PwRC and provide the

functionality to track and report on local government expenditure on PwRC.

- Take a staged, individual approach towards introducing recycled content targets for local government procurement to enable supporting tools and databases to be developed and for regional and local variations to be taken into account.
- Fund further research, development and delivery of recycling technologies and products generated from recyclables, particularly by local or regional councils.
- Establish clearing house/brokering service (virtual) matching waste/resource materials to potential users/manufacturers of PwRC to address some of the uptake barriers.
- Develop and financially support a mechanism to facilitate the generation/bulking up of volumes at an appropriate geographic distribution through cooperative arrangements across suppliers including support for infrastructure (this may be needed both at state and regional levels).
- Work with stakeholders across supply chains to find opportunities to not only reduce waste but also ensure PwRC is valued and used over and over as a renewable resource through new clean technologies.
- Improve on-site separating and processing to support improved diversion rates and reduced contamination.

3.4.3. Operation of Existing C&D Waste End-Markets in the NT

The NT is a vast federal territory in Australia located around the arid Red Centre. This territory has a tropical monsoonal climate, which features a dry and a wet season. The NT's economy is mainly concentrated around a few industries. These include construction, government and community services, and the mining sectors, which account for half of the NT's total economic output. In 2020, the construction industry employed 11,158 people, 23% down from the previous year [62]. The latest report by the ABS shows that the value of building work done in the NT has been on a sharp decline since 2014. Unlike most jurisdictions, this value has steadily shrunk from AUD $418 m in 2014 to about AUD $183 m in 2020, a 128% decrease (Figure 12). This figure contributed 6.3% [63]. to the total NT GDP in 2018–19.

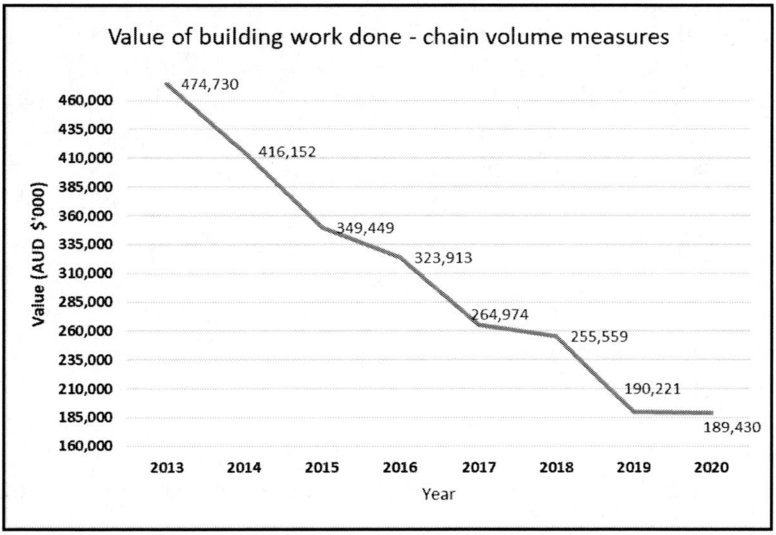

Source: ABS (2020) [34].

Figure 12. The value of construction activities in the NT 2013–2020 (until Sep).

3.4.3.1. C&D Waste Data

According to the latest report (2018–19) [2] provided by the *DAWE*, except for steel, which is largely disposed of in landfills, none of the studied waste materials are recycled in the NT (Table 31). This could be related to that national integrated data reporting system and reduced construction activities across the state.

Table 31. Priority C&D waste materials statistics in the NT

Material	Waste generation (t)	Disposal (t)	Recycling (t)
Brick	0	0	0
Concrete	0	0	0
Asphalt	0	0	0
Steel	119,811	119,811	0
Glass	94	94	0

Source: 2020 National Waste Database [36].

3.4.3.2. Existing Markets

Currently, there is no functional market for C&D waste PwRC in the NT, and much of what is generated at the C&D sites is destined for landfills. The lack of information on C&D waste management makes it difficult to examine the capacity for market development and stimulation in this territory [64].

3.4.3.2.1. Material Recovery Facilities

Shoal Bay Waste Management Facility [65]. is one of the very few centres in the *City of Darwin* that accepts a range of C&D recyclables. One division in this centre, Recycle Shop, which became operative in August 2020, sells second-hand construction materials. In Recycle Shop, which is operated by Helping People Achieve (HPA), a community-based organisation, construction industry practitioners can exchange C&D waste resources.

3.4.3.2.2. Use of PwRC in Construction Projects

In the NT, there is a limited number of projects in which the use of C&D waste PwRC has been documented. In 2020, for the first time in the NT, recycled asphalt was used to upgrade seven major roads in Darwin, through a collaboration between the City of Darwin and Downer Group. The use of PwRC in these road projects is estimated to have saved 12 t of CO_2 emission [66].

3.4.3.2.3. Government

The territory government has a pivotal role in establishing a market for PwRC. The government has begun to transition to a CE and it is expected that the territory will increase market capacity to maximise resource efficiency of construction materials. To achieve this, the government needs to develop policies encouraging promote precise and transparent waste data collection and reporting [67].

3.4.3.2.4. Recommendations

The following are the main recommendations made to boost C&D waste market development in the NT. These recommendations are extracted from experts' views published in various sources [64, 67]:

- Mandate minimum recycling content in all large-scale construction projects.
- Share recycling market information.
- Create markets for PwRC at the design and tender stages of state significant developments.
- Distribute the levy revenue to increase the capacity of recycling facilities across the state.
- Develop a communication and engagement program to ensure waste generators have a knowledge of the system they are supplying.

3.4.4. Operation of Existing C&D Waste End-Markets in Qld

Qld is an Australian state covering the continent's northeast, with a coastline stretching nearly 7,000 km. Geographically, Qld is Australia's second-largest state following WA, with a total area of 1,730,648 km². The state is the third most populated state in Australia, and the population is increasing at one of the fastest rates in the country. Based on a Population Australia estimation [68], Qld's population will reach 5.31 m by the end of June 2023. The second-largest employer in the state is the construction sector, with 245,100 people employed in the industry in 2019–20. The construction industry contributed $27.3 bn (8.1% of total output) to Qld's economy in 2019–20, making it the state's third-largest sector [69]. Since 2016, the value of building work done (all construction sectors and building types) in Qld has consistently decreased after having reached a historical peak in 2016 (Figure 13). A reduction in the waste received from interstate plus reduced construction activities have resulted in a lower amount of C&D waste generated annually, contributing to these lower figures. Furthermore, the state has one of the lowest waste recovery rates in Australia [70].

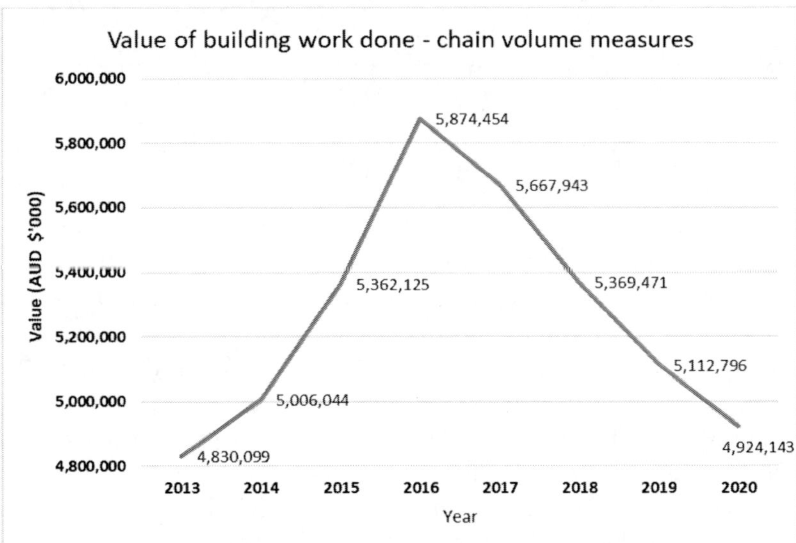

Source: ABS (2020) [34].

Figure 13. The value of construction activities in Qld 2013–2020 (until Sep).

3.4.4.1. C&D Waste Data

As outlined above, the decline in construction works and interstate waste in recent years has contributed to a fall in Qld's C&D waste volume. In 2018–19, the state's total C&D waste recovery and disposal were 3.09 mt and 2.2 mt, respectively. However, the total amount of C&D waste reported decreased by 37 kt [69] and the interstate C&D waste volume in 2018–19 declined by 19% compared to 2017–18 (1.03 mt to 864 kt). Historically, Qld's C&D waste disposal and recovery rose until 2018–19. Even still, in that year (2018–19), only about 3 mt (or 58%) of the 5.2 Mt of C&D waste reported was recovered, far behind the national average rate [69] Among the five priority materials listed in Table 32, concrete has the largest volume of waste generation and recycling in Qld. The available data [2] shows no disposal records for brick, concrete, asphalt and steel.

The state's recycling target rate was set in 2021 at 80% by 2024 [71], which is considered to be an ambitious aim to achieve in just three years.

Table 32. Priority C&D waste materials statistics in Qld

Material	Waste generation (t)	Disposal (t)	Recycling (t)
Brick	73,278	0	73,278
Concrete	1,868,779	0	1,868,779
Asphalt	438,221	0	438,221
Steel*	410,401	0	410,401
Glass*	7,502	957	6,545

Notes: * numbers are for C&D waste streams only.
Source: 2020 National Waste Database [36].

3.4.4.2. Existing Markets

The market for C&D waste is not fully developed in Qld due to several reasons including lack of incentive, insufficient recycling facilities, geographically spread cities and towns, and a low landfill levy. It is argued that a newly introduced landfill levy, currently set at $75 /t, will have a positive impact on developing end-markets for C&D waste resources [72]. in the long term. However, in the short term it can be anticipated that it might shock the waste management and resource recovery sector as they are still dealing with the consequences of the imposition of the new landfill levy [73].

Currently, there is no viable market for PwRC, and the main applications of PwRC are in public projects and local recycling facilities and retailers. However, with the recent government shift in waste management focus, the utilisation of C&D waste materials in major public projects has become mainstream and will eventually result in C&D waste end-market development throughout the state. Like other states, second-hand and C&D waste PwRC are being traded in Qld at a large scale on different platforms such as Gumtree and Facebook Marketplace.

From the priority items tabulated in Table 32, there is a potentially strong market in Qld for concrete, given the current volume of recycling. There are several facilities in Qld, such as BMI Group, Alex Fraser Group and Moreton Bay Recycling, that supply concrete recycled products including recycled concrete aggregate, drainage recycled concrete and crusher dust. The local recycling facilities do not charge waste producers for dumping concrete if they dispose of clean concrete; however, if waste producers choose to dispose of their concrete waste in a waste tip, then the charges include a $75/t landfill levy plus about $26/t for commercial waste fees [74]. Such a cost disparity—i.e., $101—is a notable financial incentive for those who are involved in concrete waste generation to consider other options.

The result of the desktop review shows that different types of recycled concrete products are being sold at various prices in Qld depending on the location, quality and pricing mechanism. Some price examples illustrating the conditions of the Qld end-market for PwRC derived from the end-of-life of this important construction material are shown in Table 33.

Table 33. PwRC price list example in Qld

Item	Bulk Landscaping Supplies* Price** (A$/t)	Moreton Bay Recycling [74] Price** (A$/t)
Crusher Dust	22–26	11
5 mm (Concrete)	35–40	N/A
10 mm (Concrete)	33–38.5	28.6
20 mm (Concrete)	28–34	17.60
40 & 70 mm (Concrete)	27–32	24.2
Road base (CBR15)	22–28	N/A

Note: *The supplier sets a minimum of 12 t order for such prices; **Prices are current as of January 2021.

Source: 2020 National Waste Database [36].

To better understand the viability of the market of a PwRC in Qld, Moreton Bay Recycling [74] compared the cost of recycled concrete versus virgin products. The comparison shows that the percentage of cost-saving can range from 3.3% (in the case of 40 mm drainage aggregate) up to 53.3% (in the case of 70–100 mm drainage aggregate). That said, in some cases (e.g., 7 mm drainage aggregate), quarry products are slightly cheaper than their recycled alternatives.

Concrete—To estimate the potential market for recycled concrete products in Qld, the following calculations are made. Assuming all recycled concrete is to be converted to 10 mm (concrete), which is priced at $28.6/t (Table 33), and accounting for 10% residual waste during recycling (Table 32), the total potential market for recycled concrete is $48.1 m/year.

Total market worth = the state's concrete recycling quantity – 10% residual waste allowance × the retail price of PwRC

Total market worth = $1,868,779 – 186,877 × $28.6 = $48,102, 397

3.4.4.2.1. Use of PwRC in Construction Projects

The history of utilisation of PwRC in construction project in Qld dates to the early 2010s, though it has not flourished until recently. With the growing popularity of the CE concept in the waste management context, the state's policymaking has transitioned towards initiatives that include maximising resource efficiency. Currently, the Qld *Department of Transport and MainRoads (TMR)* are exploring new opportunities for the use of PwRC in earthworks, drainage and concrete, focusing on diverting unwanted resources including C&D waste materials from landfill. *TMR's* trial goal is to increase the utilisation of RAP up to 40% in new asphalt and divert 8 kt of C&D waste materials from landfill. In doing so, in a collaborative construction project, Alex Fraser Group pilot-tested the application of more than 17 kt of PwRC on Brisbane's National Freight Terminal [75].

Redland City Council is also working with Alex Fraser Group and Suncoast Asphalt to resurface 1 km of road comprising significant amounts of hard plastics and RAP while cutting council costs and carbon emissions. However, the results of this trial project and their performance are yet to be publicised.

The following are two case studies in which C&D waste PwRC were used in Qld:

Case Study 1: Bruce Highway Upgrade Program [76]. – Bruce Highway is one of Australia's major highways and links to Brisbane and Cairns (Figure 14). The $8.5 billion Bruce Highway Upgrade Program is the largest road infrastructure program in Queensland. The 10-year program will improve safety, flood resilience, travel time and capacity. The road is heavily trafficked, carrying 1,500 vehicles/hour. Alex Fraser Group supplied 5,500 t of Recycled Concrete Pavement Material (RM001) for the project through civil constructors ALL ROADS. Choosing PwRC resulted in approximately 47 t of carbon savings, 1,335 t less material compared to virgin materials, 45 fewer truck movements in this project and cut the project costs by more than $20,000.

Case Study 2: Clem Jones (CLEM7) Tunnel [77] – In 2010, the construction of the Clem Jones Tunnel (CLEM7) provided a critical river crossing that bypasses the CBD to link Brisbane's growing northern and southern suburbs, with direct connections at Bowen Hills, Kangaroo Point and Woolloongabba (Figure 15). Alex Fraser Group developed a recycled concrete roadbase to engineers' specifications specifically for the project. The material was stabilised to meet the demands of the application. Recycled roadbase is approximately 15–20% lighter compared to materials produced from Qld quarried rock, which means less material is required. The benefit of using PwRC was a cost-saving of an estimated $570 k, and 725 fewer truck movements. The use of PwRC also prevented the extraction of 120 kt of natural resources, saved more than 100 kt of waste from going to landfill and reduced carbon emissions by around 1 kt.

Source: Alex Fraser (2020) [77].

Figure 14. Bruce Highway Upgrade Program; the largest road infrastructure program in Qld.

Source: Alex Fraser (2020) [78].

Figure 15. The Clem Jones Tunnel (CLEM7).

3.4.4.2.2. Government

In addition to the legislative responsibility of the government sector, the state authorities need to encourage market development through funding infrastructure expansion, such as a $100 m funding program for waste and recycling (the Resource Recovery Industry Development Program [78]), to meet the demands that are foreseeable due to a growth in landfill levy rates and diversion of waste materials. In addition to the state's latest initiatives to demonstrate the suitability of C&D waste PwRC, ongoing research to showcase the performance of infrastructure with PwRC can provide confidence in the market. Government sustainable procurement is another necessary step towards the transition to a CE in the construction industry.

3.4.4.2.3. Recommendations

The following are some state-specific recommendations for Qld to drive the creation and stimulation of end-markets for C&D waste resources. The recommendations are based on expert views provided in various sources [73]:

- Ensure the waste and resource recovery infrastructure planning is aligned with the waste strategy, market development strategy and any other upcoming regulatory decisions.
- Develop standards for materials, as well as specifications for particular products and end-markets.
- Change procurement practices to incentivise and encourage (or mandate, as appropriate) the use of PwRC where it is available.

- Review landfill levy rate annually to ensure its functionality is able to obstruct all loopholes and will be in favour of end-market development and stimulation.
- Take advantage of interstate waste transfer to provide sustainable feedstock for recycling facilities.
- Guarantee a full levy exemption will apply to residual waste materials created producing trade exposed PwRC.
- Distribute the levy revenue to increase the capacity of recycling facilities across the state.

3.4.5. Operation of Existing C&D Waste End-Markets in SA

SA, with a total land area of 983,482 km^2, is the fourth largest of Australia's states and territories by area, and fifth largest by population. It has a total of 1.77 m people [79]. It occupies one of the driest, most barren parts of the continent, but its southern fringe consists of well-watered and fertile lands and is where most of the population is located [80]. The construction industry in SA is not among the top industries; however, in terms of "industry share of output" and "gross value added (GVA)," it was ranked third among other industries in 2018–19 [81]. The value of building work done (all construction sector and types) in SA has fluctuated between 2013 and 2020 (Figure 16). Despite this, figures recorded in 2020 increased 18% from those of 2013. Defence projects are expected to make up a large percentage of the work for the Adelaide construction industry in the foreseeable future. Projects underway include the Future Frigates and Future Submarine program with the next phases of Air 7000 and Air 555 at Edinburgh in the pipeline [82]. Recently, the $25,000 HomeBuilder Grant program to recover the effects on the industry caused by COVID-19 conditions has increased construction activities and as a result, construction jobs have grown 4%, which is above the national average figures [83].

SA holds a special national and international position in the management of waste, including C&D waste materials. Unlike most places globally, waste management and resource recovery industry has been successful in collaborating with waste management responsible organisations in achieving the SA waste management strategy document [84] and a CE.

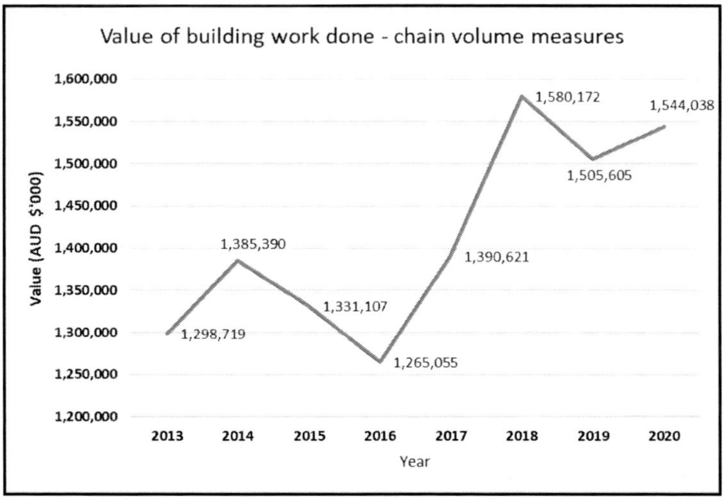

Source: ABS (2020) [34].

Figure 16. The value of construction activities in SA 2013–2020 (until Sep).

3.4.5.1. C&D Waste Data

SA continues to perform well in its efforts to recycle material and reduce landfill disposal [2]. The 2020 National Waste Report indicated that the generation of C&D waste increased by 32% from 1,231 kt in 2017 to 1,64 kt in 2019. With the increase in the C&D waste quantity, waste landfilling, energy recovery and recycling activities were also expanded. The growth rate was recorded to be 12%, 33% and 101%, respectively, reflecting the state's successful efforts to recover C&D waste materials. As presented in Table 34, for the five case study materials (i.e., brick, concrete, asphalt, steel and glass), the annual quantity of recycling outperformed disposal in 2018–19.

Table 34. Priority C&D waste materials statistics in SA

Material	Waste generation (t)	Disposal (t)	Recycling (t)
Brick	122,717	20,638	102,079
Concrete	988,757	27,355	961,402
Asphalt	285,895	69	285,826
Steel*	54,929	4,046	50,883
Glass*	331	331	0

Note: *numbers are for C&D waste streams only.
Source: 2020 National Waste Database [36].

For instance, in the same year, more than five times the amount of brick waste was recovered as was landfilled. In this state, the resource recovery rate is 85%, which ranks first in Australia. This rate consists of 80% of recycling and 4.4% energy recovery. These rates serve as the benchmark for resource recovery and the CE approach in Australia.

3.4.5.2. Existing Markets

The waste management and resource recovery sector is an important industry in SA. Leading the country in terms of waste recovery rate, SA has great potential for creating new and stimulating existing markets for certain C&D waste materials, notably brick and asphalt (Table 34). Further, SA is at the centre of waste interstate transport for recycling in Australia. Several states and territories (the NT, NSW, Vic and WA) transport their waste to SA to recycle and repurpose. For example, SA receives metal and glass from the NT and NSW [37]. The latest reports show that SA also exports some C&D waste (i.e., 1,954 t plastic recovered from C&D waste) to interstate for recycling. University of Adelaide researchers [37] have argued that the reason for such a business pattern is the financial benefit gains through waste trading, which highlights the significance of an efficient domestic market in this state. A newly published report by the state government [85]. indicated that in 2017–18 about 86% of all waste streams were recovered in the state, and the remaining were sent overseas (7%) and interstate (7%). The exported materials included mainly metals, cardboard and paper, and to a lesser extent, plastic.

A recent report by *GISA* [86]. outlines the conditions of the existing market for some C&D waste materials across this state. In SA, the resource recovery industry is a significant sector of the state's economy, with an annual turnover of about $1 bn, a contribution of more than $500 m to GSP (directly and indirectly) and employing about 4,800 people. The value of the resources recovered each year is also significant: in 2017–18, this was estimated at $356 m [86].

3.4.5.2.1. Material Recovery Facilities

Pre-cycle is an initiative that commenced in SA in 2016 [87]. The aim of this initiative is to help builders remove C&D waste materials from construction sites. The initiative outlines a six-stage process during construction. First, pre-cycle sorts recyclable materials at their source before removing them to be recycled and added to a CE in the construction industry. The operating company, Premier Insulation, recycles and repurposes the offcuts from new home construction sites, saving construction companies waste removal bills and keeping unused building materials from ending up in a landfill. Of what is collected from construction sites, 80% get used [87]; in building a typical house using this initiative, construction builders would only use one-sixth of bins and pay one-third of waste management expenses compared to conventional waste management practices. The initiative is also able to divert 30 m^3 of waste that would otherwise have been landfilled. A report by *GISA* [85] showed that this business model could increase landfill diversion to 76%, up from the 10% seen in the business-as-usual approach.

3.4.5.2.2. Use of PwRC in Construction Projects

SA is a pioneer state in using C&D waste PwRC in public infrastructure projects. Below are two examples of such an application:

Case Study 1: Built Environs, 100 Hutt Street, Adelaide (SA) [59]– 100 Hutt Street is a commercial office building and the head office of Built Environs, the national building brand of McConnell Dowell (Figure 17). The building at 100 Hutt Street was refurbished between 2007 and 2008 using the GBCA's Green Star building rating tool to demonstrate leading practice. The refurbishment achieved an overall rating of five green stars. During the construction, the project implemented a waste management plan including using 95.1% of construction waste (by weight) from the construction activities, far exceeding the Green Star credit criterion.

Case Study 2: Reclaimed asphalt as unsealed residential pavement in Regional SA [88]. – The reconstruction of Cooper Street, Kudla was undertaken by the City of Gawler's works engineering department, which chose Bitumate as the appropriate material for an unsealed surface. Cooper Street is a regional access road to five residential properties (Figure 18). Reconstruction of the road pavement in addition to the installation of side drainage was required due to regular flooding. As an unsealed surface, patrol grading is minimal, and with some residual bitumen, it provides some cohesion to reduce ravelling (particularly at the intersection) and some "re-healing" of the surface under traffic in hot weather occurs. Reclaimed asphalt is received at C&D recycling facilities predominantly in the form of slab asphalt removed from old pavements. The raw feed material is processed through crushing and screening to grading with little or no plasticity. The product is produced by Resource under the brand name "Bitumate™."

Source: Edge Environment (2012) [59].

Figure 17. Built Environs headquarter in Adelaide, SA.

Source: Realestate (2021) [89].

Figure 18. Use of Bitumate™ in a regional road in SA.

3.4.5.2.3. Government

The state government through GISA provides a range of funding to unlock the potential of the CE, develop infrastructure to process and create new products from waste, seed funding for new technologies and commercialise research in SA. Notably, the Circular Economy Market Development grant [90]. aims to encourage councils, not-for-profit organisations, research institutes and businesses that produce, manufacture, sell or promote SA-PwRC. Furthermore, the government's $190 m Recycling Modernisation Fund [91] is set to improve recycling outcomes and address critical gaps in the state recycling infrastructure. To further support efforts in developing the market at the state level, the new waste strategy document has an extensive focus on the CE. This is reflected in the fact that the waste strategy is entitled 'A Vision for a Circular Economy' [92]. Government support has progressively increased with recent changes to the accessibility of overseas end-markets. For instance, the South Australian Sustainable Procurement Working Group has been formed to maximise collaboration between all levels of government, the business community and the waste sector to develop a framework to drive end-market development for post-consumer recyclables. Such initiatives can underpin viable and sustainable markets for the state's C&D waste stream.

3.4.5.2.4. Recommendations

The following recommendations are made to increase the capacity of existing markets for C&D waste PwRC and develop new end-markets across the state. The recommendations are extracted from various sources [91] in which experts' views are considered:

- Build stable markets for remanufacturing and secondary materials.
- Increase market confidence for investments in the CE, resource recovery and waste management.
- Increase use of recycled material and build demand and markets for PwRC.
- Promote manufacturing of products and components that replace virgin materials with sustainably produced materials.
- Develop successful procurement case studies demonstrating the benefits of using recycled-content products to government and industry.
- Support the development of accredited testing for product standards and performance to increase confidence in the quality of remanufactured products.

3.4.6. Operation of Existing C&D Waste End-Markets in Tas

Tas lies about 240 km south of the state of Vic, from which it is separated by the shallow Bass Strait. The state comprises a main island called Tasmania as well as several others of its coast. The main island is roughly heart-shaped, with a maximum length and width of 320 km. The state also produces a major portion of Australia's hydroelectric power and possesses a great diversity of natural resources [93]. The state comprises a main island called Tasmania as well as several others of its coast. It has a total of 0.57 m people [80].

The construction industry in Tas has demonstrated a steadily rising trend since 2014 [34]. In 2020, the value of building work done in this state reached AUD $417 m, a 36% increase from the 2014 figure. A recent report by Inside Construction [94]. estimated that the total worth of the construction projects in 2020–21 in the state will amount to AUD $2.8 bn a year. Figure 19 illustrates the value of building work done in Tas from 2014 to 2020.

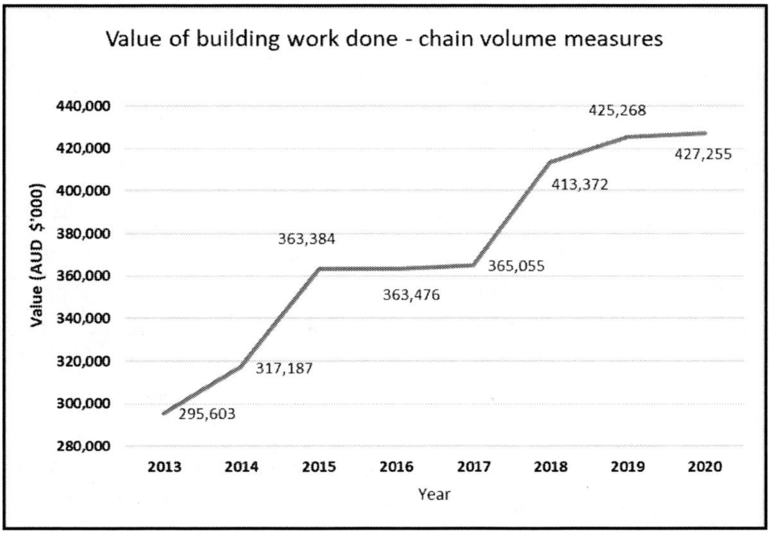

Source: ABS (2020) [34].

Figure 19. The value of construction activities in Tas 2014–2020 (until Sep).

The state has the lowest C&D waste recovery rate in Australia, with about 1% of C&D waste recovered each year. The state does not have a current waste strategy, and hence there is no official waste minimisation or recovery target; the 2019 Draft Waste Action Plan [95] targets a 40% average recovery rate for all waste streams by 2020.

3.4.6.1. C&D Waste Data

The Tas construction industry generates about 44 kt of waste annually, of which only 1% is recovered, making it the lowest rate in Australia [2] and well below the national resource recovery average. According to the latest report

(2017–18) [2] provided by the *DAWE*, none of the studied C&D waste materials is recycled across Tas (Table 35). It is most likely that these materials are landfilled as Tas does not impose a landfill levy on materials to be disposed of.

Table 35. Priority C&D waste materials statistics in Tas

Material	Waste generation (t)	Disposal (t)	Recycling (t)
Brick	0	0	0
Concrete	0	0	0
Asphalt	0	0	0
Steel	0	0	0
Glass	0	24	0

Note: *numbers are for C&D waste streams only.
Source: 2020 National Waste Database [36].

3.4.6.2. Existing Markets

In Tas there is no viable market for PwRC, due to many reasons including low construction activities, lack of landfill levy and the government policies. In a report prepared by Urban EP [96], issues were identified by waste management and resource recovery stakeholders as impediments to the creation and stimulation of the Tas market for PwRC. Among the four problem areas indicated by the three regions in Tas, insecure market for investing in waste recovery infrastructure was the major issue related to C&D waste market development. To tackle this issue, the report suggests the implementation of coordinated education, engagement, and marketing and market development measures. These two solutions involve fostering acceptance and uptake of new recovery services, stimulating demand for recovered resources, and guiding consumer and purchasing behaviour and decisions; as well as the implementation of SP. In turn, this will stimulate Tas markets for resources recovered locally and encourage the replacement of non-recyclable and single-use items with reusable content items.

While the current conditions in the market and the nature of the construction industry—in which time is money—do not favour deconstruction as opposed to demolition, Tas has shown the capacity for market development measures through private sector efforts. For instance, Hobart's Resource Work Cooperative [97] (HRWC), based in Tas and founded in 1993, carries out deconstruction that generally maximises the value of resources used in construction projects. HRWC [98] is Australia's largest not-for-profit, self-funded, workers' cooperative that aims to create employment through waste minimisation activities. The cooperative provides two services, namely urban salvage and deconstruction. Through urban salvage services, HRWC removes reusable waste directly from the demolition site. Currently, HRWC is seeking to trial this service on a larger scale. Through deconstruction, their team take down buildings by hand to recover up to 95% of materials to recover resources. HRWC partners with registered builders and a team of qualified resource workers to provide a low impact alternative with competitive rates.

3.4.6.2.1. Material Recovery Facilities

Tas's waste management system operates in three waste management group regions: Southern, Northern and Cradle Coast. Observations indicate that in each region, while there are multiple landfills, one landfill centre dominates the market for disposal services [95]. For instance, Copping Landfill takes the majority of waste in the Southern region, Launceston Waste Centre receives 84% of the waste generated in the Northern region and Dulverton Waste Management operates the biggest landfill in the Cradle Coast. In some of these landfills, C&D waste is marginally recovered.

3.4.6.2.2. Use of PwRC in Construction Projects

The use of C&D waste PwRC in Tas is minimum and limited to one project that has documented its overall information.

Case Study 1: Kingston roadworks [99] – *Kingsborough Council* applied an aggregate consisting of PwRC in roadworks in partnership with RED Group, *Close the Loop* and *Downer Group*. About 78 t of recycled asphalt (Reconophalt [100]), along with other waste resources (e.g., glass, toner and plastic bags), were used (Figure 20). The project's aim was to reseal Baynton Street, Harris Court, Lucas Street and Cleburne Street. Every tonne of Reconophalt paved on the road incorporated approximately: 526 plastic bag and packaging equivalents 250 glass bottle equivalents, toner from 18 used printer cartridges and 100kg of RAP. Roads surfaced using Reconophalt have a 15% longer life than traditional surfacing.

3.4.6.2.3. Government

While there are some attempts to use PwRC in construction projects at the local government level, the Tasmanian state government support to boost the C&D waste market is not sufficient. The Northern Tasmanian Waste Management Group (NTWMG) region has planned to provide funding to create a regional C&D recovery facility at the Launceston Waste Centre [101].

The region-based waste management groups in three councils have collaborated to publish two reference guides to help improve resource recovery in the C&D, and C&I sectors.

Source: Kingsborough Council (2020) [100].

Figure 20. Recycled roads in Kingston.

3.4.6.2.4. Recommendations

There are several recommendations that can improve the operation of the limited existing market for C&D waste across Tas. Some of these were extracted from the Feasibility Study into a Statewide Waste Management Agreement Report [95]:

- Consider CE principles as being core to its operations and a natural fit for Tas's circumstances.
- Develop strategies for priority materials.
- Implement and improve state-wide data collection, analytics and reporting.
- Establish coordinated education, engagement and marketing.
- Foster local governments engagement and procurement support to lock in demand for new services and facilitate efficient use of assets.
- Develop market development measures including SP.
- Coordinate advocacy and policy input.
- Fund infrastructure to simulate investment in recovery assets.

- Devise a market and/or statutory instrument to address gate fee differentials.
- Introduce a state-wide waste levy to encourage waste recovery and market development.

3.4.7. Operation of Existing C&D Waste End-Markets in Vic

Vic is the second most populated state in Australia. It has a total area of 227,416 km^2, which accounts for 3% of Australia's total landmass and makes it the smallest mainland state. Vic's coastline is 1,800 km long and borders the Bass Strait, the body of water separating the mainland from Tas. Based on the latest estimations, Vic's population will reach 6.8 m by the end of June 2023 [102]. Vic's construction industry employs almost 240,000 people and contributes $21.6 bn to the Victorian economy. The industry is driven by a continuing increase in Vic's population and the need to provide housing and related infrastructure [103]. The value of building work done (all construction sectors and building types) in Vic reached $9.92 bn in 2020 (Figure 21). As a result, C&D waste is the largest contributor to the state's total annual solid quantity. In 2018, compared to C&I and MSW, more C&D waste was recovered than was landfilled.

3.4.7.1. C&D Waste Data

Due to massive residential, commercial and public construction across the state, Vic's annual C&D waste generation rate is among the highest in Australia. Currently, the state's C&D waste recovery is 87%. The waste generation in the state has increased by 47% since 2017, reaching 8,133 kt in 2019. Likewise, there was a surge in C&D waste recycling and disposal, by 56% and 15%, respectively. The government's commissioned report [2] states that concrete and brick are the major C&D waste materials in Vic. Table 36 summarises waste data related to the five case study materials in Vic.

Concrete comprises the largest end-market across the state, due to its massive application in infrastructure projects and huge waste generation annually. A report prepared by *SV* [104] indicates that the end product market for this material in 2013–14 was about $40 m. This report estimated that the end product market for brick was $10 m in the same period.

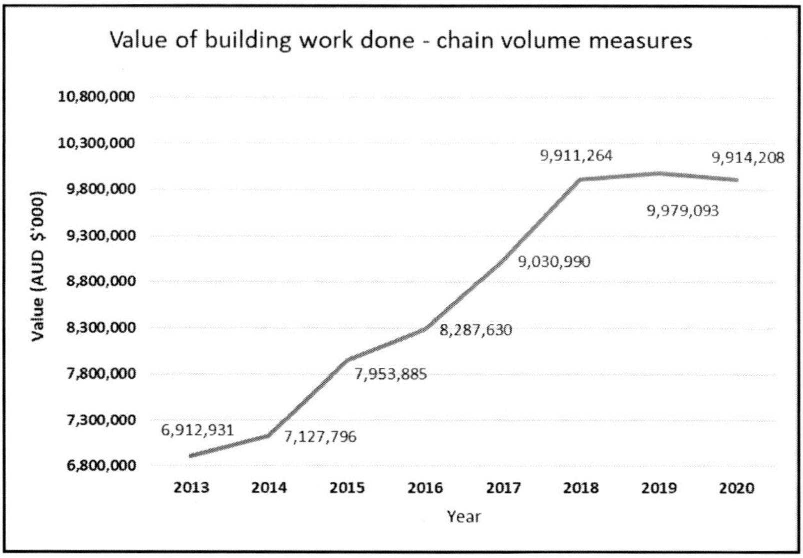

Source: ABS (2020) [34].

Figure 21. The value of construction activities in Vic 2013–2020 (until Sep).

Table 36. Priority C&D waste materials statistics in Vic

Material	Waste generation (t)	Disposal (t)	Recycling (t)
Brick	804,650	0	804,650
Concrete	2,710,460	0	2,710,460
Asphalt	231,560	0	231,560
Steel *	147,838	0	147,838
Glass*	3,331	2,339	972

Notes: *numbers are for C&D waste streams only.
Source: 2020 National Waste Database [36].

Vic transports 32,985 t of metals (25%), 2,374 t of paper & cardboard (46%), 2,074 t of plastic (74%), 445 t of TLR (45%), 75 t of glass (7%) and 656 t of organics (1%) of recovered C&D waste to interstate or overseas for reprocessing/recycling [105]. An analysis of multiple reports revealed that the major destination of this waste was probably SA [37]. It is expected that Vic's waste recycling sectors will need to adjust to meet the policy objectives announced by the *Council of Australian Governments* (COAG) in March 2020, which will progressively ban export waste from July 2020 onwards [106].

3.4.7.2. Existing Markets

Compared to other states, Vic has a strong market for PwRC. Hence, recycling facilities such as Bingo Industries and Re-purpose offer lower base prices. Furthermore, massive infrastructure projects being executed across the state provide great opportunities for stakeholders to use PwRC in construction projects. For instance, in 2019, the City of Bayside [107] used recycled asphalt in roadworks in four suburbs, in the form of 12,000 t of sustainable asphalt that contained waste materials. It is estimated that the state's waste recovery sector transition to a CE will help to create more than 3,900 new jobs and establish new skills in design, repair, efficiency and material usage across Vic [108]. Below are estimated markets for the five study materials in this study:

Concrete—assuming all recycled concrete is to be converted to 10 mm (concrete), which is typically priced at $22/t, and accounting for 10% residual waste during recycling, the total potential market for recycled concrete is $53.67 m/year.

Total market size = quantity of recycled concrete – 10% residual waste allowance × the retail price of 10mm recycled concrete aggregate [109].

Total market size = 2,710,460 t – 271,046t × $22 = $53.67 m

Brick—In Vic, a recycled brick is priced at $1.32 [110] and each brick weighs around 3.75 kg. Given the annual recycling rate (Table 36), it can be inferred that the potential market size for brick is $7.28 m. The associated calculations are provided below:

Number of recycled bricks: 804,650 t (quantity of recycled brick in WA)/3.75 kg (average weight of a brick) = 214.57 m bricks

Potential market size: 214.57 m (number of recycled bricks) – 10% residual waste allowance × $1.1 (average price of a brick) = $212.5 m

Asphalt—Recycled asphalt in Vic is worth $20/t [111] and given the state's concrete recycling quantity, which is 231,560 t (Table 36), the potential market size for this resource is $4,168,080. The calculations are as below:

Total market size = quantity of recycled asphalt – 10% residual waste allowance × the retail price of recycled product

Total market size = 231,560 t –23,156 t × $20 = $4.17 m

Steel—The potential market for steel is the largest among the study materials in Vic. Recycled stainless steel is approximately $1.26/kg [112]. Given the state's recycled steel amount (Table 36), 147,838 t, the potential market is worth $175 m (Figure 22).

Total market size = quantity of recycled steel × the retail price of recycled product

Total market size = 147,838,000 kg × $1.26/kg = $186,275,880

Figure 22. The potential market for priority materials in Vic.

3.4.7.2.1. Material Recovery Facilities

By developing guidelines, the state government supports the knowledge enhancement of resource recovery best practice management. For instance, in 2019, *SV* published a document to guide the development, management and operation of MRFs across Vic [113]. Vic's MRFs are under continuous inspection by a task force to ensure safety measures are observed. In 2014–15, Melbourne had 20 waste transfer stations, 23 resource recovery centres and 25 drop-off facilities [114]. It is estimated that through improving material efficiency and recycling under a CE, the state's economy would be boosted by $6.7 bn [107].

3.4.7.2.2. Use of PwRC in Construction Projects

Vic continues to improve the application of PwRC in large infrastructure projects. The following are some case studies demonstrating its track record of success in using PwRC projects across the state.

Case Study 1: Manufacturing with PwRC in regional Vic [115] – Integrated Recycling, located in Mildura, Vic, uses recycled plastic primarily from local agricultural sources to create durable products such as grapevine posts, landscaping products, boardwalks and railway sleepers. It also manufactures Duratrack, a recycled plastic railway sleeper that lasts more than three times longer than timber sleepers and requires less energy to manufacture than concrete or steel sleepers. This keeps waste plastic that would otherwise go to landfill at a higher value and in use for longer—a key principle of the CE.

Source: Alex Fraser (2018) [117].

Figure 23. Dingley bypass opened in 2019.

Case Study 2: Dingley Bypass, Southeast Melbourne [116] – The 6.4 km, $156 m Dingley Bypass has been completed with Alex Fraser Group materials (Figure 23). The bypass was opened in mid-March 2019, a full five months ahead of schedule. The new road makes the Alex Fraser Clarinda site, where 96% of crushed pavement materials for the project were sourced, even easier for customers to access. Throughout the two-year project, 269,000 t of Class 4 Crushed Concrete, Class 2 and 3 Crushed Concrete, and Recycled Sand was supplied. Alex Fraser Group products resulted in 23,000 t fewer used due to the density savings compared to quarried rock and 770 fewer truck movements. In addition, 1.7 kt of CO_2 have been prevented from entering the atmosphere.

3.4.7.2.3. Government

The Victorian state government's financial support for the recycling and waste management industry is wide-ranging. In 2019, the government announced a plan titled 'Recycling Victoria: A new economy', in which it will invest $300m to transform the recycling sector [107]. The fund aims to drive investment in world-class infrastructure and technology, make the state's future recycling system more sustainable, create cutting-edge local industries and support thousands of new local jobs. The state government also provides funding for projects that use PwRC [117]. The Victorian Government has released $2.6 m in grants from the Sustainable Infrastructure Fund to help roll out PwRC for local construction projects. Infrastructure projects across 79 local councils are expected to cost $8 bn over the next three years. The government intends to leverage this funding to encourage the use of PwRC.

The other market development initiative has come through *Recycling Victoria*, the Government's 10-year action plan for waste and recycling. It will invest more than $300 m to transform the state's recycling sector, reduce waste and create thousands of jobs. In 2020, *SV* launched an online directory that will feature local, Victorian PwRC. The directory will help the Victorian government source PwRC. It will be used by state and local government procurers and buyers to help them easily research, review and access PwRC.

Lastly, the government initiative to encourage localised waste management through several waste and resource recovery groups across Vic sets an example of how integrating specific conditions in the broader context of a state's waste management strategy—if properly coordinated with other organisations (e.g., *EPA*, *SV*, industry and businesses)—can drive a vital market for C&D waste PwRC. It is also noteworthy that Vic was the first Australian jurisdiction to develop a comprehensive framework that plans for waste and resource recovery infrastructure [107].

3.4.7.2.4. Recommendations

The following are some recommendations to improve the operation of Vic's existing market for PwRC. The recommendations are made according to experts' views published in various sources [112] and should emerge from collaboration between key stakeholders including the government, the construction industry and the resource recovery sector:

- Work with local and other relevant stakeholders to develop sustainable end-markets for recovered recyclables to reduce exposure to price volatility of international commodity markets.
- Recommend the use of recycled and recovered material in infrastructure design and build and the purchase of products that are made from or contain recycled or recovered material.
- Establish a framework for monitoring progress towards the CE, including the identification of indicators and metrics.

3.4.8. Operation of Existing C&D Waste End-Markets in WA

WA, the largest state in Australia, covers the entire western third of the country. It is made up mostly of the arid Outback and its total area is 2,529,875 km², which accounts for 33% of Australia's total landmass [118]. Its population is concentrated in its fertile southwest corner, home to the Margaret River wine region and the riverside capital, Perth. Based on the latest estimations, WA's population will reach 2.9 m by the end of June of 2023 [117]. WA's GSP of $285.6 bn in 2018–19 was 15% of Australia's GDP.

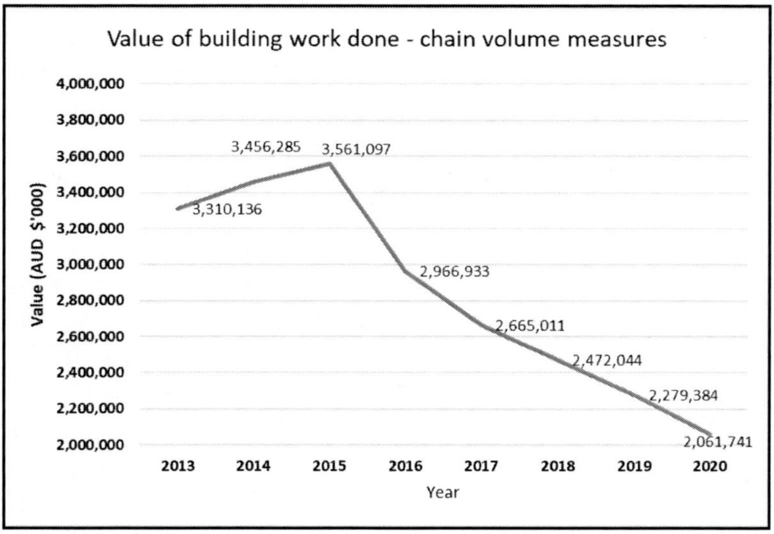

Source: ABS (2020) [34].

Figure 24. The value of construction activities in WA 2013–2020 (until Sep).

Construction is the third major contributor to the state GSP following mining, business and property services [119]. However, the latest data shows that construction activities in WA have consistently weakened since the start of 2015 in all construction sectors. In 2020, the value of building work done (all construction sectors and building types) dropped to about $2 bn (Figure 24), a 72% decrease from that of 2015, making it the largest detractor from GSP growth.34 This trend was attributed to several factors such as a lack of cash flow, poor management and the fall in mining investment [120]. Unsurprisingly, this resulted in a significant decrease in the amount of C&D waste generated in the state. In WA, C&D waste materials are considered as priority waste management resources.

3.4.8.1. C&D Waste Data

The C&D waste sector is the largest volume of waste resources diverted from landfill in WA. It is also the largest volume of waste generated of any material type; it accounts for 90% of total solid waste in the state [121]. Compared to other major states, WA features a smaller C&D waste flow [2]. In 2018–19, the state produced 1.9 mt C&D waste. The low generation of C&D waste and recycling and disposal is directly related to the state's declining trend in construction activities. According to the National Waste Report [2], the state's C&D waste recycling in 2019 increased by 26% compared to that of 2017, while C&D waste disposal saw a comparatively slight decrease (3%). C&D waste-based EfW was recorded at 2 kt. The most recent data reported by the *Waste Authority* [120] shows an 80% resource recovery rate for the C&D waste stream, which is 24% more than the average resource recovery in the state. Among the priority materials, concrete had the greatest recycling rate in WA. There is no data for waste disposal for brick, concrete, asphalt and steel available (Table 37).

There are overseas markets for scrap steel and—according to the Waste Authority report [122]—in 2018–19, more than 597 kt of non-packaging steel was destined for overseas markets (Asia), in addition to 20.5 kt that was sent interstate. The state has embarked on a few major infrastructure projects that will change the waste flow landscape. In early 2020, WA had $27.2 bn of major resource projects under construction or committed and $91.1 bn under consideration [123]. It is foreseen that these projects will act as opportunities to divert C&D waste resources from landfills by utilisation of PwRC in construction; albeit the COVID-19 pandemic situation might present some challenges.

Table 37. Priority C&D waste materials statistics in WA

Material	Waste generation (t)	Disposal (t)	Recycling (t)
Brick	19,213	0	19,213
Concrete	217,994	0	217,994
Asphalt	136,816	0	136,816
Steel*	138,870	0	138,870
Glass*	230	230	0

Note: *numbers are for C&D waste streams only.
Source: 2020 National Waste Database [36].

3.4.8.2. Existing Markets

With over 1.5 Mt of C&D materials processed annually in WA, there is a large supply of usable materials, and recently there have been some successes in developing new markets for these PwRC [124]. Currently, the largest end-market for C&D waste materials in WA is public projects. These projects have unlocked the potential of a domestic market for extensive application of PwRC. Particularly, the *Waste Authority's* initiative called 'Roads to Reuse' (RtR) [120], established in 2020, outlines the state's efforts in the use of C&D waste PwRC. The organisation's vision is to encourage state government organisations, local governments, regional councils and the private sector to use recycled C&D products in civil applications, such as road construction. It is estimated that reduced construction activities in WA, however, will cause a decline in supplying feedstock for recycling facilities and thus affect the market development in future [121]. To better understand the potential of existing markets, the following calculations for concrete, brick and asphalt are made. These are based on local prices of PwRC and the quantity of waste recycled across the state annually.

Concrete—Assuming all recycled concrete is to be converted to 10 mm (concrete), which is typically priced at $55/t, and accounting for 10% residual waste during recycling, the total potential market for recycled concrete is $63.9 m/year.

Total market size = quantity of recycled concrete – 10% residual waste allowance × the retail price of recycled product

Total market size = 217,994t – 21,800 t × $55= $10. 79 m

Brick—In WA, a recycled wire-cut red clay and recycled pressed red clay brick cost $1.31 and $1.53, respectively [125]. Each brick weighs 3.75 kg and given the annual recycling rate in WA (Table 37), it can be inferred that the potential market size for brick is $7.28 m. The associated calculations are provided below:

Number of recycled bricks: 19,213,000 kg (quantity of recycled brick in WA)/3.75 kg (average weight of a brick) = 5.13 m bricks

Potential market size: 5.13 m (number of recycled bricks × $1.42 (average price of a brick)) = $7.28 m

Asphalt—Recycled asphalt in WA is worth $9.50/t [124] and given the state's concrete recycling quantity, which is 136,816 t (Table 37), the potential market size for this resource is $63,921, 330. The calculations are as below:

Total market size = quantity of recycled asphalt – 10% residual waste allowance × the retail price of recycled product

Total market size = 136,816 t – 13, 681 t × $9.5 = $1. 17 m

Steel—The steel potential market is the largest among the study materials. The recycled stainless steel is approximately \$1.26/kg [126]. Given the state's recycled steel (Table 37), the potential market is worth \$175 m (Figure 25).

Total market size = quantity of recycled steel × the retail price of recycled product

Total market size = 138,870,000 kg × \$1.26/kg = \$174, 976,200

Figure 25. The potential market for priority materials in WA.

3.4.8.2.1. Material Recovery Facilities

WA has the largest material recovery facilities across Australia. In fact, the new Cleanway's South Guildford plant is the largest recycling plant in the southern hemisphere [127]. Before a 2019 fire incident forced it to shut down permanently, this facility accepted recyclables from 20 local governments. Meanwhile, public authorities in WA such as the *Waste Authority* and EPA attempt to increase customers' awareness of the benefits of recycling and actively support recycling businesses. For instance, a public entity, the *Southern Metropolitan Regional Council (SMRC)*, launched a directory of recycling facilities under a program called Recycle Right [128]. The program aims to encourage and assist residents and businesses in WA to recycle, reduce their waste and live more sustainably.

A survey study123 on the operation of major C&D waste recyclers in Perth indicated that:

- All materials processed for recycling and reuse comply with environmental regulations, and well and specific standards and specifications for end-use applications.
- All undertake *National Association of Testing Authorities (NATA)* testing.
- All recover, recycle or make available for reuse a range of materials and also produce a range of products.
- All have some form of in-take quality control procedures to monitoring inbound loads, including:
  - Only receiving own-loads, or known and trusted customers;
  - Visual and photo/x-ray imaging of in-bound loads;
  - Pricing to match the quality of separated/sorted materials vs. co-mingled loads; and
  - Rejecting contaminated loads (to be sent to landfill).
- All have on-site processing facilities—i.e., primary sorting and crushing, screening and handpicked sorting, secondary crushing and then stock-piling for testing as required.

The differences in these recycling facilities' operations include a focus on a particular market and product, varying approaches to separating and sorting materials, and detecting asbestos in their feedstock. The report also illustrates the main processed materials and their end-markets identified in the study WA's recycling facilities (Table 38).

Table 38. Products received, processed and relevant main end-markets

Products Received	% of total product processed	Main end-market(s)
Fill sand	High – 65% Average – 35% Low – 20%	• Civil developments and building construction • Construction blocks
Crushed recycled concrete	High – 80% Average – 35% Low – 5%	• Roads and road shoulder work • Carparks and truck parking • Driveways • Construction site and ground development • Rammed earth walls and construction blocks • Concrete batch aggregate substitute

Table 38. (Continued)

Products Received	% of total product processed	Main end-market(s)
Mixed masonry material (blended rubble – concrete, bricks, tiles, pavers)	High – 20% Average – 35% Low – 90%	• Road base • Track material • Subgrade works • Hardstands • Construction site and ground development
Recycled asphalt/Bitumen	Average ~10%	• Roads – Hard stands – Car parks
Drainage stone (crushed aggregates)	Average ~3%	• Subsoil drainage for car park runoff • Garden bed drainage
Concrete dust/fines	~2%	• Concrete batch substitute • Compacted pavements
Crushed bricks – red, cream, blended	~1%	• Ground cover for courtyard settings • Garden landscaping and placement around plants • Rockeries and around pathway stepping-stones
Brick fines – red, cream, blended	~1%	• Walkways • Verge parking areas, and driveways • Inside shade houses, and horse arenas

Source: Active Sustainability (2020) [123].

3.4.8.2.2. Use of PwRC in Construction Projects

According to industry and government reports, it makes economic sense to apply current certified C&D waste PwRC in large infrastructure projects in WA. In a research report [123] commissioned by the *Waste Authority*, C&D recyclers reported that their certified road-base products are valued at $5–6/t, while virgin quarried materials are valued at up to $20/t. Furthermore, the report indicated the figures for certified clean-fill products and their virgin alternative are $3 and $10/t. The following are some case studies in which C&D waste PwRC were used in WA.

Case Study 1: Kwinana Freeway Northbound Widening [129]. – This project was completed in 2019 and its budget was $49 m. The project constructors used approximately 25 kt of C&D waste PwRC or crushed recycled concrete (CRC) as sub-base under full-depth asphalt (Figure 26). Further, the project team provided surplus site material to smaller local projects and overall, sold 56,783 t of waste sand suitable for reuse.

Source: Building for Tomorrow (2019) [130].

Figure 26. The Kwinana Freeway northbound widening, Perth, WA.

Case Study 2: Street Pavement in Willetton, WA [131] – Boral, a major Australian construction and building materials supplier, in partnership with the *City of Canning Council*, has constructed a suburban street in Perth using four PwRC (Figure 27). The street materials include waste tyres, plastic, glass and old road pavement, all of which were originally bound for landfilling. The PwRC were combined with crushed rock and bitumen to pave a street in Willetton, WA.

Source: Boral (2019) [133].

Figure 27. Street pavement in Willetton, City of Canning council, WA.

Case Study 3: Port Coogee recycled asphalt trial [132] – The City of Cockburn, in partnership with real estate agency Frasers Property, has also paved 750 m² of the road in the waterfront community of Port Coogee, southwest of the Perth CBD (Figure 28), with a recycled product known as Reconophalt. Reconophalt comprises a variety of PwRC, including recycled plastic bags, waste toner from printer cartridges, crumb rubber from car tyres and RAP. The project used 58,000 plastic water bottles, 316 tyres and 37,500 glass beer stubbies. Densford Civil poured the Reconophalt, while the Downer Group supplied the materials. Reconophalt potentially lasts longer than standard asphalt—it has a 65% greater fatigue life.

Source: Quary (2020) [130].

Figure 28. The road pavement in the waterfront community of Port Coogee, Perth Australia.

3.4.8.2.3. Government

Public authorities are in the process of transiting to a CE and better energy efficiency. Therefore, the WA government is strongly supporting the use of PwRC, particularly in public projects. This has resulted in notable market stimulation for certain C&D waste resources such as concrete and brick. That said, the problem of large proportions of C&D waste PwRC failing to find suitable markets persists, and hence, stockpiles continue to grow [123]—currently at about 1,000,000 m^3. Indeed, stockpiling is a major issue in WA that requires immediate state government attention [134].

3.4.8.2.4. Recommendations

Some recommendations for improving the existing end-market and creating potential markets for C&D waste materials in WA are provided below. These recommendations are extracted from waste management and resource recovery experts' opinions published in different sources [123]:

- Implement uniform waste levies across the state, and ideally apply them in a manner that minimises 'border' market distortions.
- Monitor and prevent illegal waste dumping activities to encourage more material recovery.
- Develop technologies that assist authorities to identify and control illegal dumping across the state.
- Facilitate the process of obtaining licences and approvals necessary for the establishment of new facilities.
- Remedy contamination of raw materials and poor source separation.
- Increase customers' awareness of PwRC and their safety.

3.5. Barriers to Creation and Stimulation of End-Markets

As made apparent by the above analysis, across Australia, the uptake of practices related to using PwRC varies but overall is still limited, and this continues to have detrimental impacts on the environment [135].

Table 39. The main barriers to the use of PwRC

No	Barrier	Short Description
1	Increased costs of energy and transport	Sometimes recycled products are costlier than raw materials
2	Lack of knowledge on recycled products	Unfamiliarity with recycled products limits their application in construction projects
3	Limited technologies for waste recovery	Limited technologies in the waste recovery industry results in poor quality and expensive recycled products
4	Low quality, contamination and reduced performance	Less than expected quality hinders the wide application of recycled products in the industry
5	Lack of market availability of the products	Uncertainty about the market discourages the production of quality second-hand materials
6	Limitations caused by specifications, standards and permits	Specifications and standards have both positive and negative impact on the recycled product's market
7	Limited acceptability and negative perceptions (public and the industry)	Negative perceptions towards recycled products limit their usage in the industry

As summarised in Table 39, previous studies have identified the main factors affecting the use of PwRC in the construction industry. The following section provides an explanation of how these factors have hindered the application of PwRC in construction projects in Australia.

3.5.1. Increase in Energy and Transport Costs

Cost is considered to be one of the main barriers to using C&D waste PwRC in the Australian construction industry. Several studies indicate that PwRC are costlier than conventional materials [136-138]. A few reasons contribute to higher costs in C&D waste PwRC including longer transport distance [139], industrial waste sorting procedures [133], increased energy requirements for recycling, landfill levies for residual waste and high capital costs for recycling facilities [140] and equipment [141, 142]. In some circumstances, the use of PwRC is reported to be less expensive than fresh materials [59, 143]. However, if waste producers and recyclers are willing to work collaboratively, it can lead to a win-win situation for both parties [144].

3.5.2. Limited Knowledge of PwRC

There is solid evidence that a lack of knowledge of the various characteristics of PwRC is a factor that diminishes the willingness to use them in the construction industry. A lack of awareness and knowledge leads to negative market demand for PwRC. This could also lead to end-users having a negative perception of used materials, final product quality and the price of PwRC [59, 143]. Marketplaces for re-used and PwRC are yet to be mainstream, and there is still limited development of knowledge and practical impacts of these [135]. The limited communication between key stakeholders such as contractors and recyclers and a "wait and see" attitude are also affecting the limited knowledge on PwRC in the industry [133].

3.5.3. Limited Technologies for Waste Recovery

Technologies play an important role in safe and effective waste recovery. Limited technological advances and associated high capital costs in the waste recovery sector have historically hindered manufacturing commercially competitive products [136, 145]. Limited education on emerging waste recovery technologies, lack of investment in waste management technologies [146] and limited incentives for private investment in technology are key reasons affecting the uptake of recycled C&D product use.

3.5.4. Low Quality and Reduced Performance

Previous research studies indicate that construction industry stakeholders face issues such as quality, contamination and unacceptable performance when choosing PwRC over conventional materials [59, 142]. Lack of technological support, such as resources, training, limited competent staff and expertise, and the lack of a benchmarking tool [133] and systematic decontamination procedures [147] are key reasons affecting the quality and reducing the performance of C&D waste PwRC. As there is no assurance of maintaining the same quality and performance as virgin resources in the recycled product, there is a critical need for defining standards, resourcing producers and training staff members [133].

3.5.5. Lack of Market Availability of PwRC

The viability of a market has a direct mutual relationship with further use of PwRC. When there is an established market for PwRC, the economy of scale assists with further investment in recycling facilities, resulting in lower total cost, higher quality and less contamination. The imbalance of supply and demand for PwRC creates a narrow window of opportunity to trade recycled C&D products. For example, a study in Australia on recycled PVC products application reports that the cost of recycled material is the same as virgin material and recommends that the industry needs greater scale or a higher value for PVC to incentivise more PVC recycling [59]. Once these results are achieved, more industry stakeholders will be persuaded to buy and use PwRC in construction projects. Market availability is underpinned by several factors including effective supply chain, properly designed landfill levy, government incentives and subsidies, and finding new applications outside of the construction industry [135].

3.5.6. Limitations Caused by Specifications, Standards and Permits

There is an ongoing debate within the waste management and resource recovery industry about the role of specifications, standards and permits in promoting the application of PwRC in the construction industry. If the organisations need to provide indemnities, operate trials or accept risk, these also create an additional layer of complexity that may become an impediment for the uptake of C&D waste PwRC [135]. On the one hand, specifications and standards determine the quality of materials recycled, giving peace of mind to stakeholders to use these materials. On the other hand, many recyclers indicate that complicated and difficult-to-meet specifications and standards are a significant hindrance in selling PwRC to the market.

3.5.7. Limited Acceptability and Negative Perceptions (Public and the Industry)

Due to various factors, such as a lack of knowledge and a view of the products as unconventional, there is a negative perception among construction industry stakeholders towards using PwRC [135, 134]. It is important to recognise, however, that different stakeholder groups may have different perceptions [148] and they are not fixed. Views are changing as field trials show how to use these materials to their optimum performance and as virgin resources become scarcer [59]. This highlights the critical need for prototyping during design stages to test acceptability [135].

3.6. Strategies to Create and Stimulate End-Markets

There are a number of strategies that can be used to stimulate existing markets and develop new applications for C&D waste PwRC. In this review, five major clusters of enablers have been identified that can assist key stakeholders to advocate the use of PwRC in the construction industry.

3.6.1. Increase Community Awareness and Education on PwRC

Access to information and the education of construction experts are the two points of leverage for increasing the application of PwRC [149]. Therefore, bolstering efforts to inform stakeholders about the technical properties and environmental performance (primarily awarding authorities), and existing law and standards, is required. Furthermore, the distribution of more data, better documentation and reports about reference buildings assist with increasing engineers' experience with PwRC. In addition, the scientific community needs to make a more significant effort to convey their acquired knowledge to industry professionals.

3.6.2. Develop Supportive Regulations, Policies and Specifications

The government should prioritise the task of improving specific legislation and regulations, with a focus on establishing the mandatory degree of normative standards. Notably, design codes should highlight the environmental aspects of the construction and give designers more flexibility in material sourcing, leading to an increased usage of PwRC. Detailed specifications should also be developed in the form of application-specific quality standards for different PwRC as another option to increase the demand for such products. The introduction of mandatory policies (unified, stakeholder-oriented policies) [133]—such as SP to meet specific reuse/recycling targets, more purchasing power and grants that are offered to companies to help initiate recycling locally—would encourage both recyclers and builders to invest in waste management solutions [150].

3.6.3. Facilitate Sustainability Rating Programs

Another promising route is the implementation of sustainability rating programs that suggest the use of PwRC in construction projects. Some reports [151, 152] have demonstrated that running these programs leads to further adoption of PwRC in the construction industry. However, there is evidence proving otherwise [153]. Among others, it seems that the design of rating systems, developers' biases and lack of incentives to motive end-users are the most influential reasons for their ineffectiveness. As such, scholarly and political engagement with business practice is urgently needed to facilitate effective sustainability programs [154].

3.6.4. Promote Recycled Product Certification

A recycled product certification that is awarded following material testing, plus quality control, can increase the adoption of PwRC in the construction industry. As different supply sources will cause variation in quality, stricter quality controls for PwRC are required [155]. A government agency should be established to control the quality and production of these products. Guidance on clear technical specifications or standards on the use of PwRC for structural applications will enable recyclers and producers to adhere to industry standards [133]. This will provide industry stakeholders the confidence in these products needed to commit to their uptake in the long term.

3.6.5. Advocate Targeted Technologies and Innovative Practice

The waste management and resource recovery industry should embrace new technologies and innovation. Technological advances help PwRC producers meet market demands through products having reasonable price and quality. Notably, governments, as the main stakeholder responsible for protecting the environment, should support the industry by providing a fund to invest in technologies and innovation [156]. For instance, Building Information Modelling (BIM) can produce information on the exact quantities to order or prepare, resulting in minimum generation. Another example of technologies and innovative practice include eBins that are equipped with compact cameras, programmable and compatible with BIM applications, can make waste transportation better coordinated, more straightforward and cost-effective [157].

Conclusion

This review demonstrates differences and similarities in the existing end-markets' operation for C&D waste resources across Australian states and territories. For each state and territory, several recommendations were provided to assist policymakers, researchers and practitioners to contribute to the creation and stimulation of successful end-markets. Furthermore, this chapter identified major barriers to the operation of existing C&D waste end-markets, which included an increase in energy and transport costs, limited knowledge of PwRC, limited technologies for waste recovery, low quality and reduced performance, lack of market availability of the PwRC, limitations caused by complex specifications and standards, limited acceptability and negative perceptions. To overcome these barriers, several strategies were put forward, namely, increase community awareness, develop supportive regulations, promote sustainability programs, promote product certifications, and advocate targeted technologies and innovations. These barriers and strategies can form a management framework that informs decisions, policies and best practice management in the AEC industry to move towards a CE.

There were a few limitations that presented challenges in data collection. Below is a selection of these limitations, along with the strategies used to minimise their impact on the study's quality.

Firstly, in the absences of primary data, this chapter is confined to secondary data that is publicly available. Some of the resources were outdated and might not reflect the current market conditions. Furthermore, due to using desktop research, the resources found were not necessarily consistent in terms of the methodology applied. Therefore, a degree of information disparity is to be expected. That said, care was taken to ensure that the most recent studies, reports and resources that were also consistent are used where available.

Secondly, the information sensitivity of construction and resource recovery centres hindered the collection of data pertaining to the estimation of market size and strength for the priority materials. For instance, as the market is highly competitive, recycling facilities, suppliers and those who are involved in trading C&D waste PwRC do not tend to reveal their price list on their webpages, and the final price is typically determined after reviewing order size, transport distance and other available quotes. To this end, price averaging was applied to all prices obtained from various available resources to ensure a true reflection of the market for different PwRC under study.

Thirdly, the availability of waste data in the smaller states and territories (e.g., the NT) was another barrier. While it is possible that there is a market for C&D waste PwRC in these places, the lack of documented information impedes our ability to capture market operation in these areas. To minimise the impact of this, the public authorities responsible for waste management in these states and territories were contacted for more information, although this strategy did not always result in the required information being acquired. Lastly, there is a caveat regarding the use of the national waste dataset: it does not provide complete data for the waste fates of the priority case materials in each category and territory. Therefore, where available, supplementary data were sought from jurisdictional reports and datasets.

Chapter 4

Impact of Sustainable Procurement on the Operation of C&D Waste End-Markets

4.1. Introduction

For years, economic performance has been the dominant factor in the practice of public purchasing. However, with pressure on natural resources and increased public awareness of the adverse effects of human activities on the environment, governments and the corporate world now consider sustainability as a factor in the procurement of goods and services. This approach, Sustainable Procurement (SP) integrates requirements, specifications and criteria that are compatible with the protection of the environment and the society.

SP is defined by various organisations for different settings and applications. The definition accepted by the UN, the UK government and the *Australasian Procurement and Construction Council (APCC)* is:

> "A process whereby organisations meet their needs for goods, services, works and utilities in a way that achieves value for money on a whole life basis in terms of generating benefits not only to the organisation, but also to society and the economy, whilst minimising damage to the environment" [152].

Other variations of SP include Green Public Procurement (GPP), Sustainable Public Procurement (SPP) [158], Environmental Purchasing (EP), Circular Procurement (CP) and Environmentally Preferable Procurement (EPP). Table 40 presents variations of SP schemes and their definitions.

SP builds on the notions of a CE and cradle-to-cradle design that are measured through different tools including lifecycle assessment techniques. This policy approach "*is consistent with the principles of sustainable development, such as ensuring a strong, healthy and just society, living within environmental limits, and promoting good governance*" [159]. Some examples of SP practice in Australia are provided in Section 4.3.

Table 40. Variations of sustainable procurement schemes and their definitions

Concept	Definition	Source
Sustainable sourcing	Managing all aspects of the upstream elements of the supply chain to maximise the triple bottom line which entails the social, economic, and environmental performance.	Pagell and Shevchenko (2014) [16]
Green public procurement	Procuring goods, services and works with a reduced environmental impact throughout their lifecycle when compared to goods, services and works with the same primary function that would otherwise be produced.	Diófási and Valkó (2014) [161]
Circular Procurement	A process that enables the purchasing party to ensure that, at the end of their service life or useful life, products or materials will be re-used effectively in a new cycle, where the products and materials crucially retain their quality and value. Circular procurement is the process in which a product, a service or a project is purchased according to the principles of a CE. In this process the technical aspects of the product are as circular as possible, taking maintenance and return policies at the end of the use period into account, as well as including financial incentives to guarantee circular use	PIANOo (2020) [162] Van Oppen et al. (2018) [163]
Environmental Purchasing	The inclusion of environmental factors in decisions on the purchase of products and/or services	Australian Government (2003) [164]
Environmentally Preferable Procurement	involves purchasing products or services that have a lesser or reduced effect on human health and the environment when compared with competing products or services that serve the same purpose.	NCDENR (2010) [165]

Several benefits of SP application have been identified in the literature. For instance, Pick (2017) reported benefits such as stimulating the local economy, contributing to the brand of 'Zero Waste' and other sustainability goals and reducing municipal operating costs by selecting durable and reusable materials [166]. In the context of the construction industry, the implementation of SP can provide environmental benefits through the usage of recycled construction materials. For instance, SP could result in a decline in greenhouse gas emissions (GHG) by reducing energy consumption. Craighill and Powell (1999) estimated that about 80% of all energy required to produce a building is used in the production and transport of materials [167]. Overall, the adoption of SP in the construction industry helps to create demand for PwRC with a minimum environmental footprint including in terms of water and energy.

4.1.1. Review of Sustainable Procurement Application: Global Context

An analysis of SP policies, practices and experiences in the global context identifies potential motivations for those states and organisations seeking to establish CE and achieve sustainable outcomes. At the national and international levels, research indicate that there has been an increase in the application of this policy approach, albeit under various names. For instance, in South Korea, one study shows that all public institutions' total expenditure on green products increased from USD 759 m in 2006 to USD 2,945 m in 2017 [168]. In addition, various reports indicate that the extent to which this policy approach's associated benefits are being realised has expanded globally. For instance, in the EU, SP represents 16% of the member countries' GDP, while in the Organisation for Economic Cooperation and Development (OECD) countries, it varies between 8% and 25% [169]. In the US, the Environmentally Preferable Purchasing (EPP) program has generated strong cost and environmental benefits, including the purchase of nearly 7 m Electronic Product Environmental Assessment Tool (EPEAT)-registered products by the federal government in 2018, resulting in cost savings of around $182.5 m [170]. As a global leader in public procurement practices, China launched Green Public Procurement (GPP) in 2004. However, the literature reported that unclear regulatory requirements, costs associated with GPP implementation and procurement professionals' low awareness of GPP benefits have hindered GPP expansion in China [171, 172].

In the context of the AEC industry, the literature provides scant information on SP performance, particularly in the area of waste diversion from landfills. Multiple research publications describe the implementation of SP without quantitative evidence of its efficacy [173, 174]. For instance, in France, by including GPP into its Energy Transition for Green Growth Act 2015 [175], the French government aimed to boost the use of recycled aggregate (RA) in road construction projects by 60% by 2020. In Finland, the BE's sustainable procurement target is set at 29% [176]. Some SP policies in South Korea have encouraged the acquisition of certified PwRC, particularly in the AEC industry [165]. In Canada (Vancouver), research from the business sector indicates that a significant number of Canadian municipalities have begun implementing SP policies for civil construction projects [177]. However, follow-up research revealed that the implementation of SP in Vancouver has mostly fallen short of its anticipated objectives due to a lack of funding. [163, 178]. In Malaysia, the AEC industry 's implementation of SP is in its infancy and acceptance is dispersed fragmented [179]. In China, SP mandating the use of at least 15% PwRC in public projects is anticipated to result in the reuse of more than 1.1 m m3 of reclaimed bricks and 1 Mt of CRC [180].

4.1.2. Sustainable Procurement in Australia: The Need for Research

Australia has witnessed a significant rise in C&D waste output and recycling in the building and engineering sector [2]. SP must be the focal point of waste management planning in the Australian built environment to ensure minimal waste landfilling. For this reason, it must be properly understood, which includes identifying potential hurdles and enablers, studying its realised impacts and relevant activities, and analysing its local and national regulation. This approach is recommended in numerous Australian waste-related recommendations and strategy documents, including the National Waste Report [2]. Yet there remains a dearth of scientific study on SP in the built environment.

While in recent years, this problem has been the subject of a small number of research projects, these were limited, either in focus or the chosen research methodology. Specifically, such studies focused on carbon emission reductions [181], COVID-19 [182], certain stakeholders [183], and market development [184]. Because these studies used research methods such as case studies, meaning they have constrained generalisability [185], they therefore offer limited information regarding SP in the AEC industry. Notably, an earlier literature review [153] found that the relationship between procurement and C&D waste minimisation in Australia has not been thoroughly investigated. Further, multiple studies [186, 187] have demonstrated the need to evaluate the effect of procurement strategies on waste management planning. Moreover, the exposure of previous researchers to the real-time concerns stated by stakeholders in their research projects has revealed deficiencies in the Australian AEC industry's CE and waste management policy and practise. Survey analysis [188], a series of interviews, specialised workshops and industry seminars, and other personal communications have captured the concerns of stakeholders and alerted researchers in this field to significant gaps in dealing with this issue. In light of this, the current study aims to fill this gap and provide a deeper and less limited understanding of this policy approach.

The review findings are expected to contribute to the theory and practice of SP and policy reforms in the AEC industry. This research is timely as the Australian federal government is seeking to maximise the benefits of SP nationally. Lastly, the AEC industry itself has embarked on a journey to gradually adopt a CE [189], which in part requires knowledge of SP. Notably, this research is a part of a wider national research project that aims to enhance the growth of the C&D waste stream's waste market.

This chapter focuses on how the use of PwRC in Australia can be increased by employing a SP strategy. The aim is to examine the current condition of SP in the AEC industry's waste management and then develop a model to maximise its implementation. The following objectives were established to accomplish the research aim:

1) To identify SP principles to guide policy development.
2) To explore potential barriers to and enablers of SP in the AEC industry.
3) To analyse existing national and territorial rules, policies, and guidelines that support SP.
4) To analyse SP's practical outcomes.
5) To develop a model to facilitate the successful implementation of SP in the AEC industry.

4.2. Approach

4.2.1. Research Design

A mixed-method approach was adopted to address the five study objectives using qualitative and quantitative secondary data. To apply this research approach, multiple data collection methods suitable for each objective were applied. Using multiple data collection methods in construction management research is proven to offer a wide range of benefits [190, 191] chiefly, the provision of more substantial evidence and more confidence in research findings. The findings from the first four objectives were used to create a model to guide efforts to implement SP in the Australian AEC industry (objective 5).

4.2.2. Data Collection and Analysis

A literature review method was employed to understand SP principles (Objective 1), challenges and drivers (Objective 2). The primary data sources included academic literature, industry and government reports, and other non-conventional resources such as company websites and project catalogues. The keywords used in the desktop research were 'sustainable procurement', 'green procurement' [and other alternatives], 'government purchasing power', 'built environment', 'C&D waste', 'CE' and 'the construction industry'. In Australia, encouraging the use of PwRC through SP guidelines is a strategy only recently introduced, and its practical application does not have a long history. Therefore, it was expected that there would be little evidence demonstrating its suitability in the Australian context with regards to the operation and size of C&D waste end-markets. Consequently, this study first seeks to understand how this waste management strategy has impacted overseas C&D waste end-markets that are more established and for which there is more substantial evidence. As per Objective 2, an analysis of the international and Australian studies was conducted to understand the main challenges and enablers of SP in the worldwide experience. The criteria for the selected resources for this analysis were:

- Academic literature written in English and published within the last decade (2013–2022);
- Policies and guidelines that are currently in effect;
- Research scope defined only for the AEC industry; and
- Findings based on empirical evidence including an analysis of Australian construction projects completed in the last two decades (2003–2022).

Next, from the limited data available, case studies were selected and analysed to determine the practical outcomes of application of SP in the local context (Objective 4). Two groups of case studies were used to address this objective. Firstly, four recently established case studies (i.e., Buy Recycled, ASPIRE, ecologiQ and Waste Forum) were selected as successful examples of government-funded initiatives aiming to improve the operation of SP and enhance resource circularity in two Australian states, i.e., Vic and WA. Secondly, the utilisation of PwRC across the country was assessed using various construction project cases.

For Objective 3, policy analysis was utilised to demonstrate how the Australian policy environment supports the implementation of this policy approach in the AEC industry. For the policy analysis, the most recent versions of national and state SP guidelines, waste strategy documents and pertinent policies were considered.

In terms of data analysis, a variety of analytical models were utilised. The collected literature was subjected to content analysis to identify the most representative SP principles for the Australian context and to identify the significant challenges and drivers. The content analysis approach was adopted to analyse national and local government documents. Lastly, descriptive analysis was applied to demonstrate how PwRC were utilised in various Australian construction projects.

4.3. Results

4.3.1. Descriptive findings

In total, 49 pieces of literature were identified that meet the selection criteria mentioned above. These include 27 government documents, 11 journal articles, six industry and private organisations' reports, two conference papers, two research theses and one book chapter. More than 64% of the literature was published within the last five years.

4.3.2. Sustainable Procurement Significance, Benefits and Principles in the Australian Context

The concept of SP has gained the interest of the leading players in the BE industry, according to anecdotal evidence. This section provides a summary of the opinions of Australians regarding the significance, benefits and guiding principles of SP for C&D waste management. The involvement of government agencies in the broader application of SP practices through the development of specifications, accreditation, quality assurance and raising awareness of the implications of the PwRC (i.e., financial, social, ethical and environmental) and services would be advantageous for boosting the market demand for recovered C&D materials [192].

According to the results of a survey study [181], this policy approach, investment in technology and infrastructure, and landfill levies are three important factors that have a substantial impact on PwRC market development. Another study [193] reveals that the AEC industry's stakeholders regard SP as a crucial solution for organisational waste management if it is a viable choice or the client agrees to pay more. In a recent investigation [180], identified eight major areas in which this policy's benefits can be realised. These include strategic analysis, target formulation, task assignment, planning support, support for tendering, consistency maintenance, operation support and post-review. In 2020, the Australian government issued a new edition of the Sustainable Procurement Guideline, succeeding the 2010 version [194]. This document provides a framework for the Australian government to build on efforts to improve sustainability outcomes and mainstream sustainability principles in future procurement. It provides a list of benefits to the purchaser (government), market, society and the environment that is achieved through buying PwRC. Figure 29 integrates this list with the model proposed by the Australian ISO 20400 Committee [195] to offer a comprehensive overview of the benefits of SP as perceived by the Australian government.

To determine the fundamental principles and functions of SP in the Australian context, the relevant literature was analysed. The conclusion of the analysis demonstrates that SP standards are theoretically supported on 11 principles (Figure 30). These principles pertain to many facets of SP and can serve as a guide for public and private organisations to build their own SP strategies. These principles are described in Table 41, and for each principle, a suggested action (in the final column) is provided to transform the principle into SP of PwRC in the BE context.

4.3.3. Barriers

Barriers play an important role in moulding SP practices and the variation in the outcomes of SP. The literature has uncovered various factors that hinder the promotion and application of SP initiatives. The nature of these issues is heavily context-specific, though there could be similarities among different contexts. Table 42 summarises studies investigating SP barriers in developed and developing nations and lists 30 barriers identified in the literature. The result of the analysis indicates that the two major barriers are 'the lack of supportive organisational culture' and 'uncertainty about the PwRC quality'.

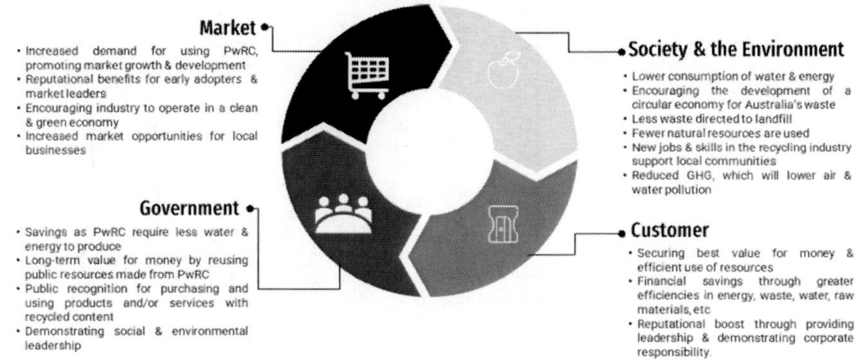

Source: Adapted from Australian Government (2020) and Australian ISO 20400 Committee (2018) [190, 191].

Figure 29. The perceived benefits of sustainable procurement by government, market, society and the environment, and customers.

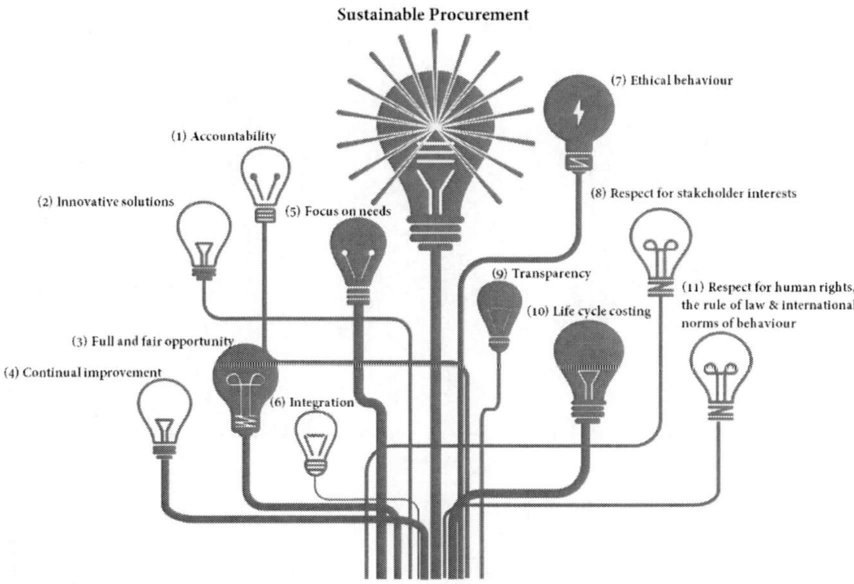

Source: Adapted from Local Government NSW (2021) [196].

Figure 30. Primary principles of sustainable procurement.

Table 41. The functionality of principles of sustainable procurement and their relevance to PwRC procurement

No	Principle	Description	Action relevant to PwRC procurement
1	Accountability	Be accountable for its impacts on society, the economy and the environment, including the impacts of the organisation's supply chain	To increase the price of using virgin resources in order to hold end-users accountable for their procurement policies and practises
2	Innovative solutions	Seek solutions to address its sustainability objectives and encourage innovative procurement practices to promote more sustainable outcomes throughout the entire supply chain	To design policies that promote innovation in the acquisition of PwRC, hence increasing their adoption
3	Full and fair opportunity	Avoid bias, and prejudice in all procurement decision-making	To prevent PwRC from being compared to virgin materials
4	Continual improvement	Work towards continually improving its sustainability practices and outcomes, and encouraging its supply chains to do the same	To optimise the AEC industry's supply chain by integrating PwRC into procurement strategy and processes
5	Focus on needs	Review demand, buy only what is needed and seek more sustainable alternatives first	To prioritise PwRC in material specifications for low-value applications
6	Integration	Ensure that sustainability is integrated into all existing procurement practices to maximise sustainable outcomes	To encourage the AEC industry to increase its capability for sustainable procurement
7	Ethical behaviour	Behave ethically and promote ethical behaviour throughout its supply chains	To monitor waste handling and processing activities to reduce the possibility of acquiring PwRC
8	Respect for stakeholder interests	Respect, consider and respond to the interests of stakeholders impacted by its procurement activities	To provide a level playing field for all stakeholders involved in the use of PwRC
9	Transparency	Be transparent about its procurement processes and how its decisions and activities impact the environment, society and the economy	To maintain transparency in handling, processing and procurement of PwRC
10	Lifecycle costing	Consider the cost incurred, the value for money achieved and also the costs and benefits on society, the environment and the economy resulting from its procurement activities	To consider the social and environmental advantages of acquiring PwRC
11	Respect for human rights, the rule of law and international norms of behaviour	Be aware of any violations throughout its supply chains and actively encourage its suppliers to do the same	To lessen the demand for virgin resources by procuring PwRC

Source: Adapted from Local Government NSW (2021) [192].

Table 42. Summary of barriers to applying sustainable procurement in the AEC industry

Reference		1	2	3	4	5	6	7	8	9	10	11	12	13	14	15	16	17	18	19	20	21	22	23	24	25	26	27	28	29	30
	Study context	Lack of long-term planning	Higher costs of PwRC	Lack of methods to measure sustainability	Lack of knowledge about sustainable products	Obligation to procure from local suppliers to lower costs and keep the quality	Unavailability of certain PwRC	Lack of government support	Misconception about PwRC production cost & time	Lack of motivation	Lack of supportive organisational culture	Lack of inclusion of PwRC in construction specification documents	Lack of mandatory imposition	Uncertainty about PwRC quality	Lack of operations	Lack of an organisational policy	Struggle to procurement behavioural change	Lack of market & market barriers	Traditional procurement procedures	Oversupply of PwRC	Lack of sustainable suppliers	Time pressure to complete projects, minimising the possibility of sustainable procurement	Difficulty in assessing the non-environmental impact	The complexity of cost-benefit analysis for PwRC	Lack of training and education for suppliers	Involvement of various stakeholders	Shortage of relevant resources	Suppliers focus on price than quality	Desire to diversify contractors	Economic uncertainty	Lack of aptitude for sustainable procurement by top management
Da Costa and Da Motta (2019) [166]	Brazil	◆	◆		◆	◆		◆			◆														◆						
Delmonico et al. (2018) [197][J]	Brazil									◆	◆				◆			◆												◆	
Pick (2017) [163]	Canada											◆		◆						◆		◆				◆		◆	◆		
Bougrain (2017) [15]	France									◆	◆			◆	◆																
McMurray et al. (2014) [198]	Malaysia	◆			◆			◆																			◆				
Islam et al. (2017) [199]	Saudi Arabia										◆																				◆

Table 42. (Continued)

Reference		1	2	3	4	5	6	7	8	9	10	11	12	13	14	15	16	17	18	19	20	21	22	23	24	25	26	27	28	29	30
	Study context	Lack of long-term planning	Higher costs of PwRC	Lack of methods to measure sustainability	Lack of knowledge about sustainable products	Obligation to procure from local suppliers to lower costs and keep the quality	Unavailability of certain PwRC	Lack of government support	Misconception about PwRC production cost & time	Lack of motivation	Lack of supportive organisational culture	Lack of inclusion of PwRC in construction specification documents	Lack of mandatory imposition	Uncertainty about PwRC quality	Lack of operations	Lack of an organisational policy	Struggle to procurement behavioural change	Lack of market & market barriers	Traditional procurement procedures	Oversupply of PwRC	Lack of sustainable suppliers	Time pressure to complete projects, minimising the possibility of sustainable procurement	Difficulty in assessing the non-environmental impact	The complexity of cost-benefit analysis for PwRC	Lack of training and education for suppliers	Involvement of various stakeholders	Shortage of relevant resources	Suppliers focus on price than quality	Desire to diversify contractors	Economic uncertainty	Lack of aptitude for sustainable procurement by top management
Adell and Schaefe (2019) [200]	South Korea					◆	◆							◆																	
Hasselbalch et al. (2015) [203]	UN		◆										◆			◆		◆	◆												
United Nations Environmental Programme (2016) [201]	World-wide								◆								◆				◆		◆	◆							

Table 43. Summary of enablers of applying sustainable procurement in the AEC industry

Reference	Study context	1	2	3	4	5	6	7	8	9	10	11	12	13	14	15	16	17	18	19	20
		Developing clear & supportive regulations	Project owners' commitment to sustainability	Project team capability	Strategic partnership	Customer expectation/pressure	Providing incentives & rewards	Strong technical expertise	Quality assurance to maintain the consistency	Corporate reputation	Proper planning & anticipating the arrangement	Improving sustainability knowledge through education	Setting a measurable target for using PwRC	Maintaining transparency & good governance	Mandatory application of sustainable procurement	Conducting a cost-benefit analysis to educate project leaders & investors	Potential to export locally manufactured goods	Screening out suppliers with a lack of commitment to sustainability during the tendering stage	Knowledge sharing among project stakeholders	Effective coordination to address challenges & conflicts	Procurement post-review
Shooshtarian et al. (2022) [202]	Australia								◆		◆		◆	◆		◆		◆		◆	◆
Ershadi et al. (2021) [180]	Australia																				
Pick (2017) [163]	Canada		◆											◆							
Zhu et al. (2013) [203]	China	◆				◆	◆														
Bougrain (2017) [15]	France												◆								
Bohari et al. (2017) [176]	Malaysia	◆	◆		◆		◆					◆			◆				◆		
Sajjad et al. (2020) [204]	New Zealand					◆				◆											
Ogunsanya et al. (2019) [205]	Nigeria	◆										◆		◆							
Gounden (2016) [206]	South Africa	◆			◆			◆									◆				

4.3.4. Enablers

Experts in procurement suggest that SP planning is contingent on the fulfilment of a series of mostly context-specific factors. Therefore, this study explored the main enablers of SP. As shown in Table 43, previous research has identified a variety of enabling factors, primarily in the public sector of developed and developing nations. Of the 20 enablers identified in this review, 'implementing clear and supporting regulations' and 'maintaining transparency and good governance' are the two most important.

4.3.5. Conceptual Model for Implementation and Promotion of Sustainable Procurement

A conceptual model to implement and promote SP can guide SP practices and be followed by procurement experts in both the public and private sectors. In this section, some of these models that have previously been developed in different contextual conditions are outlined. These models are not necessarily specific to C&D waste but have relevance and application to this waste stream.

4.3.5.1. GPP Conceptual Model in China

This model is underpinned by three clusters of factors including pressures or drivers, knowledge and GPP practices. External pressure such as regulations can stimulate GPP practices while officials need tools and knowledge to effectively implement GPP practices. In this model (Figure 31), developers posit that various drivers and pressures can motivate the adoption of GPP practices while GPP knowledge moderates this motivation. GPP practices can be defined as integrating environmental concerns into public procurement processes. Thus, the GPP practices constructed in this model include green products and service preference, green logistics and green supplier monitoring. Drivers can derive from regulations, non-governmental stakeholders, and rewards and incentive gains. Knowledge, as a construct, can include knowledge about detailed regulatory requirements, responsibilities, and experiences of developed country practices.

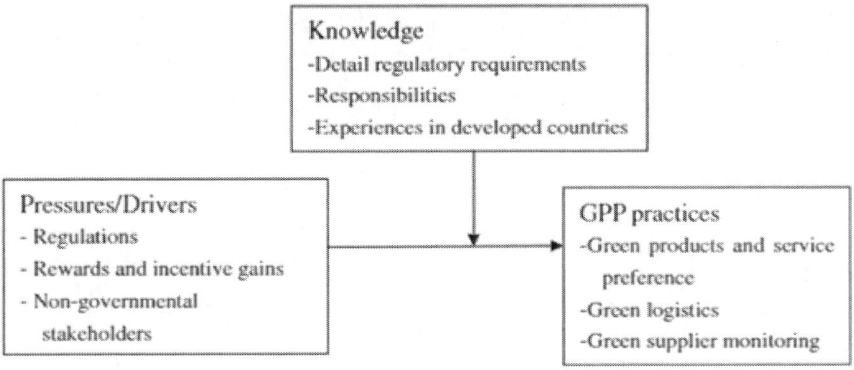

Source: Zhu et al. (2013) [199].

Figure 31. A conceptual model developed to promote GPP in China.

4.3.5.2. Sequential Implementation Model

This model is constructed in alignment with barriers identified the study of Hasselbalch et al., (2015) [207] in which a series of interviews about SP implementation and promotion were conducted with experts employed in UN agencies. This model highlights which higher-order issues must be resolved before the lower ones can be operationalised effectively (Figure 32). For instance, training staff in SP makes little sense before securing the political mandate to do SP and institutionalising this mandate in the organisation's strategies and policies.

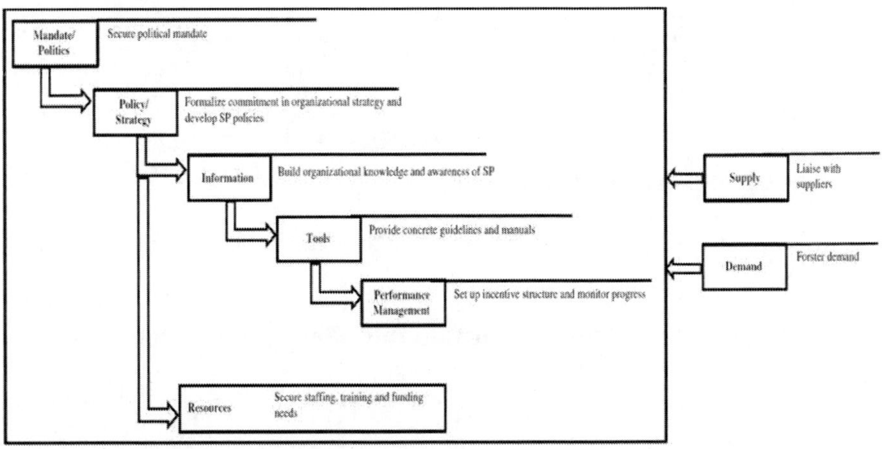

Source: Hasselbalch et al. (2015) [203].

Figure 32. Sustainable procurement sequential implementation model.

4.3.5.3. Resource Efficient Business Models (REBus) [208]

When adopting circular thinking, the procurement cycle can be proactively used to influence key areas of the construction process, from product design and manufacture, to use and disposal. The inner loop of Figure 33 shows a simplified process cycle for construction projects. It highlights the key stages for embedding circular thinking and decisions within the product and procurement cycles. Each stage has differing degrees of complexity depending on the nature of the construction project, e.g., building or infrastructure. A key element of circular thinking in procurement is embedding this thinking and action by the relevant stakeholders in each stage of the cycle.

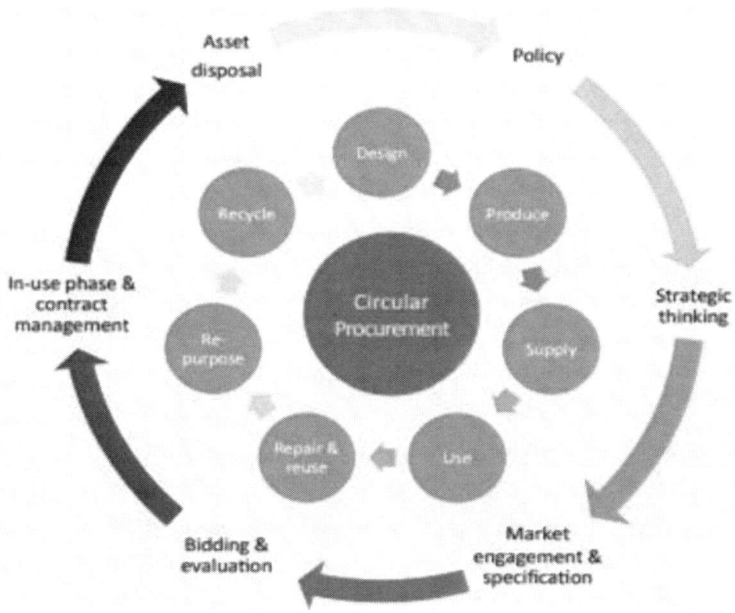

Source: CP REBus model [206].

Figure 33. Embedding circular thinking within the material and procurement cycles.

4.3.6. Review of the Existing Sustainable Procurement Policies in Australia

As described in Chapter 1, waste regulation in Australia occurs at the state level. However, certain policies of national concern, such as SP policies, are developed at the national level. In this section, policy analysis is therefore undertaken at both the national and state levels. In general, strategies for SP produced by various governmental bodies are based on a combination of considerations. These include the necessity to future-proof themselves largely in relation to supply constraints, the capacity to meet the demand of emerging markets, cost pressures and the capacity to lower these through energy usage and waste reduction [209]. This section examines the procurement policies and principles governing SP in Australia.

4.3.6.1. National Guidelines and Policies

Regarding waste management, numerous national policies and guidelines have emphasised the environmental impact of government buying practices (Table 44). For instance, the National Strategy for Ecologically Sustainable Development (ESD) [210], the NWP (2018) and other SP policies significantly support the reuse of PwRC.

Table 44. Sustainable procurement in national guidelines and policies

Guideline policy	Issuing organisation	Increased usage of PwRC through sustainable procurement
2020 Sustainable procurement Guide [190]	*Australian Government-DAWE*	Strategy 8 (sustainable procurement by governments) and Strategy 9 (sustainable procurement by businesses and individuals)
2019 NWP Action Plan [35]	*Australian Government*	Under Target 4 of the NWP Action Plan, all levels of government and industry have committed to significantly increasing their use of PwRC
National Waste Policy (2018) [23]	*Australian Government*	The Australian government has committed to considering environmental sustainability when purchasing goods and services
Environment Protection and Biodiversity Conservation Act 1999 [212]	*Australian Government*	Australian government agencies to include information about their performance on Ecologically Sustainable Development principles in their Annual Reports
Environmental Purchasing Guide 2003 [160]	*Australian Government*	EP can achieve several benefits, such as reducing waste (which can reduce waste disposal costs)

The Australian government established its first SP policy in 2003 [160]. This policy establishes the key areas for government agencies to consider when evaluating the environmental implications of their procurement activities. However, the guideline provides no instructions regarding the procurement of PwRC. The policy was revised in 2020 [190], and a five-step approach for integrating sustainability into the procurement processes of Australian organisations were introduced. These steps are as follows: (1) plan the procurement (identify the need, identify the sustainability outcomes, assess the risks and opportunities, and engage in market research and engagement); (2) approach the market (specify sustainability requirements, develop key performance indicators and establish sustainability evaluation criteria); (3) evaluate and engage (assess tender responses, debrief unsuccessful tenderers); and (4) report and manage (monitor sustainable compliance, performance) (identify opportunities to improve, monitor and track progress, and share your experience). In 2013, the federal government published a report on SP that provides an overview of the Australian government's procurement policies and operations in 2010–11 and 2011–12 that embody SP concepts and practises [211].

The main NWP, published in 2018, proposes a set of strategies to reduce waste generation through avoidance, reduction, recycling and reuse of waste resources [23]. Notably, the policy emphasises the importance of using the concepts of a CE to promote the innovative and recurrent use of existing resources. For this to happen, this policy prioritises two strategies relating to improved SP in Australia: Strategy 8 (government SP) and Strategy 9 (SP by businesses and individuals). These two strategies encourage public and private organisations to focus on the increased procurement of PwRC. In addition, through the 2019 NWP Action Plan [35], the federal government defines a number of critical initiatives to achieve the waste minimisation targets, including a recommendation to strengthen governments' purchasing power for recycling. The Action Plan presupposes that SP can increase demand for PwRC relative to virgin materials. Furthermore, it fosters innovation and investment in recycling to fulfil the need of emerging markets, supports local jobs and businesses by preserving the value of PwRC and promotes behaviour change across the economy. In addition to the policies and standards listed above, each government agency has internal policies that promote conscientious procurement concerning resource efficiency and the use of PwRC. Further to each of the policies and guidelines tabulated above, each government organisation has internal policies that encourage conscious procurement involving themes such as resource efficiency and the use of PwRC.

4.3.6.2. Jurisdictional Guidelines, Policies and Waste Strategy Documents

Of the Australian state and territory governments, the ACT, NSW, SA, Qld, Vic and WA have developed a SP guideline. These guidelines are linked with several other jurisdictions' regulations and policies. For instance, the Local Government Act 1993 (NSW), Local Government (General) Regulation 2005 (NSW), NSW Procurement Policy Framework for NSW Government Agencies, Tendering Guidelines for NSW Government 2009, and Local Council's Policies and Vision Statements all provide legal support for the implementation of the Sustainable Procurement Guide: for local government in NSW [192]. Furthermore, procurement of PwRC should comply with material specifications developed and administered by public agencies in each state/territory. Among the jurisdictional policies, the Victorian Recycled First Policy 2021 is the most relevant driver for using PwRC. Table 45 summarises the jurisdictional SP guidelines and their primary objectives.

This section analyses the jurisdictional waste management strategy documents (WMDs) to provide an overview of each state and territory's government regarding the implementation of SP of PwRC and the development of end-markets. Some state and territory WMDs highlight that SP can play a crucial part in the waste management of Australian jurisdictions. A word count analysis was used to evaluate the position of SP in WMDs in different states and territories. The three keywords used for this analysis were 'sustainable procurement', 'government procurement' and 'purchasing power'. The analysis suggests that less than 2% of study SMDs content referenced SP. Among the states and territories, WA and SA had the largest frequency of the study keywords; these two documents were recently published, i.e., in 2020 and 2021, respectively. Currently, Vic and Tas do not possess a WMD. Tas instead produced a Draft Waste Action Plan in 2019 to reinforce waste management practices throughout the state, but it does not mention the influence of SP on waste management. SA, Qld, and WA's WMDs place the most importance on SP in terms of market development. Table 46 is a summary of statements regarding the consideration of SP in WMDs that result in the creation of demand for PwRC.

Table 45. Sustainable procurement in state and territory guidelines and policies

State/Territory	Guideline policy	Issuing organisation	Policy objective/s
ACT	Sustainable Procurement Policy 2015 [213]	*ACT Government*	Advises that waste should be looked at as a resource opportunity where products can be re-introduced into another product lifecycle.
NSW	Sustainable Procurement Guide: for local government in NSW [192]	*Local Government NSW*	Provide general advice on how to procure goods and services sustainably
SA	SA Sustainable Procurement Guideline [214]	*SA State Government*	Urges the state procurement experts to consider both the end-of-life disposal of assets and also the waste produced by these products and services throughout their life.
Vic	Social Procurement Framework Recycled First	*Vic Government*	Social Procurement Framework: Incorporates sustainable procurement practices and includes requirements of PwRC, waste management and energy consumption for (government agencies) buyers. Recycled First policy: Outlines that all bidders on major transport projects will be required to demonstrate how they will optimise recycled or reused Vic's materials. Provides the government with data on PwRC for a better understanding of the supply chain.
Qld	Integrating sustainability into the procurement process [215]·		Highlights the opportunities and strategies that exist to address environmental and social impacts during procurement planning; during supplier engagement; and through the management of supply arrangements, including measurement and reporting.
WA	Guide to Sustainable Procurement 2017 [216]	*Department of Finance*	Seeks to reduce waste and by-products (e.g., waste avoidance, reuse, use of PwRC, recycling and resource recovery).

Table 46. Sustainable procurement in state and territory waste strategy documents

State/Territory	Strategy document	Sustainable procurement	Count [Per.]. of reference
ACT	ACT Waste Management Strategy 2011-2025 [217]	• Government procurement is recommended as a strategy (2.6) to recover resources fully • The *ACT Government* can use sustainable procurement principles to provide a market driver for increased use of PwRC in the goods and works that it procures. It can also encourage service suppliers to PwRC where practical • The *ACT Government* will review the specifications used for government tendering to identify where recyclable alternatives can replace non-recyclable materials, for example, in the tendering of construction and landscaping projects • As part of its proposed Carbon Neutral ACT Government Framework, the *ACT Government* will pursue additional sustainable procurement practices	12 [0.08%]
NT	Waste Management Strategy for the Northern Territory 2015–2022 [218]	• No mention of sustainable procurement	0 [0%]
NSW	NSW's Waste and Sustainable Materials Strategy 2041 [219]	• Indicates the opportunities to increase government procurement of local PwRC • The government will report annually on the use of PwRC in government procurement and its associated impact on emissions and waste reduction • The government will publish a directory of PwRC suppliers and a register of upcoming government infrastructure and construction projects that will procure PwRC	6 [.05%]
Qld	Waste Management and Resource Recovery Strategy [220]	• Under Strategic priority 3 (building economic opportunity), one government action is to consider how procurement can stimulate demand for PwRC manufactured in Qld • The *Qld Government* will work with local government and the waste management and resource recovery sector to develop a consistent procurement contract framework for waste management and resource recovery services • Local governments should support the *Qld Government* by adopting national or state standards for PwRC in procurement to help stimulate demand for products containing PwRC	4 [0.08%]

Table 46. (Continued)

State/Territory	Strategy document	Sustainable procurement	Count [Per.]. of reference
SA	SA's Waste Strategy 2020-20257	• Investigate barriers to sustainable procurement and identify measures to overcome these • Investigate financial instruments, policies and other measures that would aid in providing a level playing field for local manufacturing of PwRC • Advocate for state government and local government adoption of sustainable procurement and 'buy-back' policies • Identify and recommend priority recovered materials and PwRC to be mandated for use in the government and industry procurement system • Develop successful procurement case studies demonstrating the benefits of using PwRC • Identify relevant training needs for procurement practitioners and develop tools for capacity building in sustainable procurement • Collaborate in and advocate for nationally consistent standards and/or frameworks for the requirement of PwRC in government procurement • Support the development of accredited testing for product standards and performance to increase confidence in the quality of remanufactured products • Develop government fit-out requirements to support increased resource recovery and material reuse and repurpose • Ensure a robust regulatory environment that supports local market development for PwRC • Include PwRC measures in government infrastructure projects • Develop monitoring and reporting mechanisms for sustainable procurement • Develop strategies to develop economically viable products from PwRC	22 [0.1%]

State/Territory	Strategy document	Sustainable procurement	Count [Per.]. of reference
Tas	Draft Waste Action Plan 2019 [221]	Boost demand for PwRC through the adoption of sustainable procurement practices across State and local government	3 [0.04%]
Vic	NA	NA	NA
WA	WA Waste Avoidance and Resource Recovery Strategy 2030 [222]	• The document places an increased focus on promoting procurement decisions that prefer local markets and play a role in supporting the development of a remanufacturing industry within WA • Local, state and commonwealth governments can influence, educate and inform, and be significant consumers whose purchasing decisions and procurement policies can have very positive impacts and influence. They have important legislative and regulatory roles and develop and implement strategies. Australia is also part of global action on waste management. • Headline strategy: Implement sustainable government procurement practices that encourage greater use of PwRC and support local market development	17 [0.13%]

4.3.7. Public State Agencies Involved in Sustainable Procurement of PwRC

In each state and territory, various public organisations are directly or indirectly engaged in the procurement of PwRC in construction projects. Figure 34 presents an analysis of the state agencies that are responsible for the effective use of PwRC in construction projects. As is evident, Vic and WA have the largest number of agencies involved in the utilisation of these products.

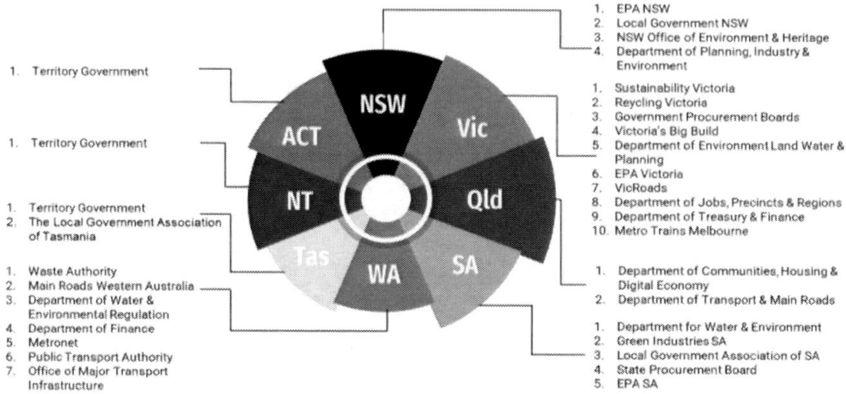

Figure 34. State and territory public agencies are currently involved in the sustainable procurement of PwRC.

4.3.8. Sustainable Procurement Initiatives in Australia

Some states in Australia have begun identifying suppliers of PwRC in order to assist government organisations in identifying SP options. Current initiatives in Vic and WA aiming to increase the use of PwRC through SP include the following.

4.3.8.1. Buy Recycled

In 2020, *Recycling Victoria* was established as a waste regulatory body to strengthen the state's waste and recycling sector [223]. *Recycling Victoria*, in collaboration with *SV*, launched an online directory (Buy Recycled) that features local PwRC and helps the *Vic Government* and public to procure them. Notably, the initiative supports the state and local government procurers to have better access to available PwRC across the state. To ensure successful implementation of PwRC SP, this initiative requires each supplier to meet the following five requirements:

1) Products must have some amount of recycled content, but otherwise, there is no minimum PwRC requirement.
2) Products must be certified against any claims made. For example, a third-party Environmental Product Declaration (EPD).
3) Products must meet all legislative and regulatory requirements.
4) Products must meet obligations regarding environmental claims under the ACCC Competition and Consumer Act 2010.
5) Any claims must meet the ACCC Green marketing and the Australian Consumer Law guidelines.

Furthermore, Recycling Victoria aims to promote the use of PwRC in the state's transportation infrastructure projects by implementing the Recycled First Policy [224]. ecologiQ is commissioned to support the objectives of the Recycled First Policy and the Buy Recycled initiative to overhaul the state's recycling sector. Also, ecologiQ has developed PwRC reference guides for road and rail-specific applications and ancillary infrastructure. The guides are an ideal starting point for information about different types of PwRC available in Vic and the current specifications for their use.

4.3.8.2. Advisory System for Processing, Innovation and Resource Exchange (ASPIRE)

ASPIRE is another initiative related to SP. Developed by the *Commonwealth Scientific and Industrial Research Organisation (CSIRO)* under the Vic's State Government's Digital Futures Fund in partnership with several Vic's councils, its operation officially commenced in 2018 [225]. ASPIRE intelligently matches businesses with potential purchasers or recyclers of waste by-products. This system requires patrons to enter details about the type and quantity of their exchangeable inputs and waste materials (outputs). ASPIRE's Supply Chain Options Model uses this data to determine optimal sources and destinations for the materials, including options for aggregation with other local businesses, and appropriate investment opportunities such as compactors for low-density waste materials and local recyclers. ASPIRE is deployed using existing established council and manufacturing business networks and supports local government business sustainability programs. It captures and codifies small-to-medium enterprise (SME) material inputs, outputs (waste and by-products) and processes and has a powerful optimisation model that takes this data and provides an SME user with: (a) suggested business-to-business (B2B) resource matches, both substitute inputs or sources and output destinations, (b) personalised search results to support the suggested matches and (c) case studies for related resource matches.

4.3.8.3. Roads to Reuse (RtR)

In WA, the Waste Avoidance and Resource Recovery Strategy [226]· governs the evolution of the C&D waste sector including defining waste management actions. The actions are operationalised in a platform called Waste Forum, which is a multidisciplinary sustainability organisation that was established by *MainRoads WA* in 2018 with a focus on innovation and collaboration. The Forum includes WA *DWER*, *Waste Authority*, and *MainRoads WA* representatives, as well as waste and recycling sector specialists. The Forum encourages sustainable purchasing in WA [227]· through supporting innovative initiatives such as the RtR program. The RtR program is the result of a collaboration between the WA *Waste Authority*, the *DWER* and WA local governments, and it intends to enhance the use of PwRC in state-wide public projects by financing pilot projects. Table 47 displays the utilisation of CRC in road construction in WA.

In addition, Les Merchant, Manager of Materials Engineering Pavement and Surfacing for MainRoads WA, reported that approximately 65 kt of crushed glass, 34 kt of CRC, 2.4 kt of crumbed rubber and 20 kt of rubber were utilised in WA infrastructure developments in 2020 as a result of this project [228]. Intriguingly, the context enables the latter group to control the entire process to guarantee that impending obstacles, as indicated by them, are effectively addressed. The Waste Forum provides a platform for government and business to communicate in order to develop the capacity to decide on ways to review product specifications for PwRC. The platform unites recycling businesses to market and make the product specifications known to all parties participating in the CRC supply chain. Periodically, in partnership with the network's industry, the forum arranges training courses to educate stakeholders on the requirements of PwRC. This networking has allowed government organisations to engage with other waste groups, such as the *Infrastructure Sustainability Council* (ISC) WA Working group, to gain a more holistic understanding of the issues related to SP.

Table 47. Statistics on the use of CRC in civil construction projects facilitated by the RtR program

Project	Original Target	CRC (T) 2019/20	CRC (T) 2020/21	Forecast CRC
Armadale Road to Northlake Road Bridge	17,355		29,014	0
High St/Stirling Hwy	12,000		12,236	0
Murdoch Drive Connection	4,000	7,287	0	0
Kwinana Freeway Northbound Widening	16,000	24,263	0	0
Leach - Welshpool	8,900	0	2,351	6,549
Karel Avenue	0	3,332	857	0
Tonkin Gap	20,000		3,531	20,000
Total		34,882	46,132	

Source: MainRoads WA (cited in Shooshtarian et al. 2022) [229].

4.3.9. Evaluation of Sustainable Procurement

This section evaluates the impact of SP policies and initiatives by examining the quantities of PwRC utilised in Australian construction projects. Multiple pilot studies and actual projects are currently underway in Australia; however, the results of the application of PwRC have not yet been documented. Therefore, this analysis only includes completed or almost finished projects. Currently, the largest market for PwRC is infrastructure projects, where the government's purchasing power is leveraged to increase the use of PwRC. Table 48 provides a summary of important case studies with reportable environmental benefits in the domain of public infrastructure.

4.4. Emergent Model for Sustainable Procurement in the AEC Industry

4.4.1. Model Development and Recommendations

To promote the implementation of SP practices in the AEC industry, an effective model was developed, known as the Emergent Model for Sustainable Procurement (Figure 35). This model draws upon extensive research on the key factors that enable or hinder SP adoption, as well as an evaluation of relevant policy frameworks. To ensure successful implementation of this model in organizational procurement processes, active engagement and collaboration among various stakeholders is essential.

Table 48. Quantitative summary of sustainable procurement induced utilisation of PwRC in various infrastructure projects

Project specifications			Sustainability outcome	Ref.
Location	Name	Duration		
Melbourne, Vic	Dingley Bypass	2014–2016	269 kt of crushed concrete and recycled sand were supplied throughout the two-year project. The application of these products resulted in 23 fewer kt used due to the density savings compared to quarried rock and 770 fewer truck movements. In addition, 1.6 kt of CO_2 has been prevented from entering the atmosphere.	Alex Fraser Group (2016) [230]
Melbourne, Vic	The Kororoit Creek road level crossing removal	2017–2018	The project used 1.3 kt of recycled glass sand at the project sites. The cost of the recycled glass sand used in this project was about half the price of virgin material due to shorter transport distances. It is also safer to handle, as it presents a lower respiratory hazard than traditional sand.	*Vic Government (2019)* [231]
Perth, WA	Armadale Road to Northlake Road Bridge	2019–2021	This project that was facilitated by the R2R program used 29 kt of CRC, which equates to almost one-third of the permanent road subbase coming from PwRC Furthermore, 14 kt of fill material was imported from nearby projects, resulting in a 10% increase of reused fill. More than 5% of all asphalt laid incorporated reclaimed asphalt materials, totalling 2.5 kt of PwRC in asphalt pavement. All retaining walls were constructed from Eco-Blocks, which are local PwRC that have been crushed, graded and repurposed.	Cited in Shooshtarian et al. (2022) [224]
Sydney, NSW	Paramatta Light Rail	2019–present	More than 15,000 m of a single rail, 13,650 rail sleepers, 13 km of overhead wires and the track ballast from a closed rail line were used to construct this project. The PwRC will provide approximately 30% of the track required along the 12 km light rail route.	*NSW Government (2022)* [232]

As depicted in Figure 35, the four factors mentioned above are the building blocks of the proposed model. The proposed model relates to five stakeholder groups: policymakers, owners of recycling facilities, construction managers, end-users (public and private sectors), and education and research providers. Depending on the circumstances, some or all of the components of this model can be utilised by other nations to boost the circularity of their AEC industry.

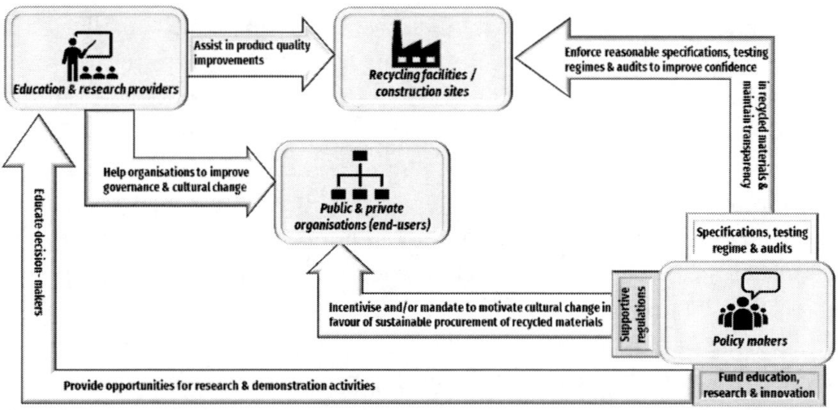

Figure 35. The emergent model for sustainable procurement in the AEC industry.

As stated previously, the analysis revealed that the two major barriers are 'the lack of supportive organisational culture' and 'uncertainty about PwRC quality'; whereas the two major enablers are 'developing clear and supportive regulations' and 'maintaining transparency and good governance'. In the AEC industry, the lack of a supportive organisational culture can diminish the efforts made to procure PwRC. Due to the novelty of the usage of PwRC in certain applications, technical and environmental concerns may diminish the desirability of these materials. Furthermore, new ideas are often not easily ingrained in an organisation characterised by decades of stagnation [233]. Therefore, it is essential that these companies' leaders (decision-makers) participate in the organisation's cultural transformation process, as they are responsible for selecting the approach to implement SP [234]. Organisational transformation for a supportive environment is driven by future thinking and circularity that are based on education of key stakeholders.

Many research studies have indicated that the uncertainty around the quality of PwRC is a major barrier to their wide procurement. This issue might partially be linked with low advances in manufacturing PwRC, as indicated by Bougrain (2017) in the case of recycled concrete in France [15]. One study in Australia [235], meanwhile, reports that the BE experts believed that some PwRC could not easily be used as new materials and sometimes there is no guarantee of their quality. Thus, government funding is necessary to encourage meaningful research and innovation to build high-quality PwRC that can be demonstrated in pilot construction projects.

Further, many of the proposed strategies for the effective execution of PwRC's SP depend on the existence of clear and supportive regulations. It is often argued that unsupportive regulations, combined with complicated and difficult to achieve standards and materials specifications, are further reasons that discourage the procurement of PwRC in construction projects [146, 184]. For this reason, the policymakers play a crucial role in ensuring that regulations are in favour of procurement of PwRC.

Moreover, to instil confidence among PwRC end-users, it is crucial for the set-up of any SP practice and policy to prioritise maximum transparency. This can be achieved through effective governance practices in both the public and private sectors.

Drawing on this model, the following recommendations are provided for the improvement of the application of SP policy and practice in the Australian AEC industry. These recommendations highlight the important role of policymakers and authorities to encourage different stakeholders to consider SP:

- Government organisations, such as Sustainability Victoria in Vic, should provide funding opportunities to reinforce research and innovation (R&I) activities that would result in increase the quality of PwRC.
- Local governments should foster collaboration with research institutes, such as CSIRO and universities, to carry out demonstration projects exhibiting the performance and reliability of PwRC to end-users.
- Education authorities such as the *Tertiary Education Quality and Standards Agency (TEQSA)* should stimulate educational programs to build the knowledge, skills and capabilities necessary to develop resilient supply chains optimised for the increased use of PwRC by end-users.

- Regulatory bodies, such as state EPAs, should modify policy frameworks that encourage or mandate (sustainable) procurement of PwRC in public and private construction projects.
- Government funded industry associations, such as *ISC* and *Green Building Council of Australia (GBCA)* should create a reasonable set-up for specifying, testing and auditing PwRC that will result in the production of consistent and acceptable material quality.
- State governments should guide waste producers and recycling facilities to provide transparency in their waste material handling and processing.

4.4.2. Research Contribution

This work adds to the development of informed procurement policies, the theory of SP and the practice of waste management in the AEC industry. According to the findings, the AEC industry lacks informed policies with the capacity to encourage players to implement SP practices. The model proposed and analytical findings establish the groundwork for policy adjustments with measurable sustainable outcomes. As the first review study conducted in Australia, the study also increases local knowledge and awareness of a crucial CE implementation strategy in the AEC industry. This contribution is made by identifying the significant challenges and enablers of SP, as well as exploring its guiding principles and benefits. Overall, the research in this chapter reveals the modifications necessary for the sector to optimise its procurement methods in favour of PwRC.

Conclusion

In the Australian AEC industry, the issue of C&D waste has become a cause of concern. SP as a policy strategy, among other techniques, can alter the status quo of the C&D waste management system and aid the Australian AEC industry in adopting a CE. The purpose of this review was to assess the current condition of SP in the Australian AEC industry. The study adopted a mixed-method research approach based on existing data that partially reflect the current state of PwRC procurement factors affecting sustainability. These primarily consist of the Australian policy framework at both the national and state levels, existing SP initiatives, the evidence for the use of PwRC through SP, as well as the main enablers and hurdles.

The findings reveal that the two most significant obstacles to SP are the lack of supportive organisational culture and uncertainty about PwRC quality. In addition, implementing clear and supporting regulations and maintaining transparency and good governance are identified as the two most important enablers. Furthermore, the study developed a model that can assist stakeholders to find opportunities for adopting SP principles at the policy formulation, education and procurement stages. To achieve this objective, a set of recommendations is provided to foster collaboration between the five key stakeholders (policymakers, research and education providers, end-users, recycling facility owners and project managers in the AEC industry) to increase the likelihood of PwRC in construction project material procurement.

The primary constraint of the review conducted in this chapter is the lack of secondary data, which prevents a comprehensive evaluation of the effectiveness of SP in the BE industry. To mitigate the impact of this constraint, a comprehensive analysis of SP evidence in the BE was done. This enquiry focused on academic literature, government and industry reports, and pertinent policies. Future studies should investigate the attitudes of key stakeholders regarding the use of SP in order to uncover perspectives pertaining to SP's practical implementation, including how stakeholders strike a balance between competing interests, to make this approach standard in the sector's procurement planning and practise.

Chapter 5

Impacts of COVID-19 on the C&D Waste Management and Resource Recovery Industry

5.1. Introduction

The COVID-19 pandemic has created significant challenges to many sectors, including housing, building and infrastructure, leading to many alterations and amendments to waste management practices [236]. In Australia, the first case of the novel coronavirus was reported on 19 January 2020 [237], and the number of cases has continued to rise ever since. The Australian government's COVID-19 statistics show that, as of 22 September 2021, Australia had recorded 90,372 total cases. While Australia has been one of the few successful countries in containing the disease, it was estimated to be one of the most affected economies in the Australasia region. The predictions suggested a shrink of 6.7% in the Australian economy in 2020 [238]. As a result, many industries have faced many financial challenges; the waste management and resource recovery industry is no exception. In the first years of the pandemic, containment of the spread of COVID-19 pandemic through limitations on commercial activities, mobility, construction and manufacturing sectors significantly affected waste management operations [239]. Specifically, the C&D waste management and recovery industry were impacted by the pandemic through a change in waste amount, composition, timing/frequency, distribution and risk, which affected handling and treatment practices [240].

5.1.1. Literature Review

In recent decades, Australian states including NSW, Vic, SA and Qld have experienced a significant volume of construction activities to meet the needs of a growing population and to keep up with economic growth [34]. During the initial COVID-19 pandemic period, the government economic recovery plan, including a homebuilder program, led to a surge of new developments, particularly residential housing. Within this context, the management of the ensuing C&D waste has become a major challenge [241]. The construction industry is accountable for 44% of the solid waste generated every year in Australia [2], and Australia is among the top 10 countries in the OECD for producing solid waste. For example, Australia produced a total of 27 Mt of C&D waste in 2018–19 [2]. This situation emphasises the critical need for better waste management practices to divert waste from landfills.185 Due to growing C&D waste quantities, scarcity of landfills and the long-term adverse environmental, economic and social consequences of C&D waste disposal, sustainable C&D waste management is needed for the protection of public health and natural ecosystems [242]. This need—for comprehensive and integrated waste management systems, enabling technologies, rating schemes and policies—is becoming broadly appreciated.

The key barriers to C&D waste management have been identified as the lack of advanced technological capabilities and knowledge, the mechanisation of demolition works, the geographic spread of products to be returned for reuse or recycling, inconsistent levy rates across states, the quality of the recycled material and attitudes and behaviour towards waste management practices [188]. The COVID-19 challenge exacerbates some of these barriers and has added another layer of complexity to the management the large volumes of C&D waste generated in Australia, posing additional challenges to waste operators, recyclers, and other relevant industry and government decision-makers.

Observations of industry performance worldwide suggest that industry operators have been impacted by unprecedented difficulties due to the pandemic [243-245]. In 2020, for example, the United States of America (U.S) ceased recycling programs in some of their cities, as authorities became concerned about the risk of COVID-19 spreading in recycling centres [246]. A survey by the *US National Waste & Recycling Association* (2020) attempted to discover the impact on the industry by inviting operators to express how their operations were disrupted by the pandemic [247]. The study revealed that, in 2020, the primary impacts on the industry were employee safety and health concerns, shifts in operational best practices, and changes in revenues, services, waste streams and waste volumes. The survey also showed that waste hauliers were the most affected operators, staffing was the main short-term impact, and two-thirds of participants indicated moderate to severe operations impacts due to COVID-19 restrictions. Another survey, facilitated by *Waste Today* magazine [241], showed that around 30% of respondents operating in the US waste management industry were ordered to close their facilities by state officials or had voluntarily idled one or more facilities. The survey results specific to C&D waste streams indicated that 50% of inbound flow had been affected from moderately to severely. Other reported impacts included 32% cases of layoffs (a further 21% were considering layoffs) and around 66% of businesses putting one or more purchases on hold. Love and Rieland (2020) state that given worker safety concerns, low market prices for scrap materials, a slowed economy and cheaper alternatives for disposal, many communities and businesses across the US had temporarily suspended the collection of recyclables during pandemic [248].

With disruptions in waste recovery activities, the environment is under increased threat worldwide [242]. In some UK rural communities, for example, disrupted services increased waste mismanagement by 300% [249] Roberts et al. (2020) reported that with fewer options available, traditional waste management practices such as landfills and incineration replaced more sustainable measures such as recycling, with adverse effects on the environment [250]. Government organisations sometimes exacerbate the issue. For instance, the UK Environment Agency allowed temporary waste and incineration ash storage at sites, which further threatened the environment. In Australia, waste-related industry associations have played a leading role to inform adaptive policies and adjusting to COVID-19 conditions. The Australian Council of Recycling (ACOR) surveyed the pandemic conditions in 2020 [251] and suggested that governments should consider their findings as part of the ongoing response to COVID-19 and more broadly. Meanwhile, the *Waste Management Resource Recovery Association of Australia (WMRR)* continues to update the industry on COVID-19 related policies and measures in different jurisdictions.

5.1.2. Research Aim and Scope

There has been limited research that provides deep insight into the impact of the COVID-19 pandemic on the industry and its responses. Therefore, this chapter asks the following research question: What are the impacts of COVID-19 on the C&D waste management and resource recovery industry? It aims to explore the COVID-19 pandemic impacts on the Australian C&D waste recycling and construction industry and the practical responses enacted to overcome pandemic-induced challenges. The following sections present the research approach and key findings.

5.2. Approach

To understand the impacts of the COVID-19 pandemic on the Australian C&D waste management and resource Recovery industry, first the background context to the topic is analysed through the existing literature and then the research questions are defined as the preliminary step of the exploratory study [179, 252]. This approach enabled the authors to evaluate available evidence and provide groundwork for more rigorous studies in the future [253].

In Australia, industry activities are regulated at the jurisdictional level of the nation's six states and two territories. One public organisation in each jurisdiction is responsible for governing waste-related activities [254] and these organisations have produced information regarding the impact of the COVID-19 pandemic on the industry. Therefore, where possible, findings on the impact are explicitly provided for each jurisdiction.

5.2.1. Data Collection

A purposive sampling strategy was employed to recruit a wide range of participants across the C&D waste supply chain representing five Australian jurisdictions (WA, Qld, Vic, NSW and SA). In 2018–2019, these five states generated 26.2 million tonnes altogether (>97% of the total C&D waste in Australia), of which 6.03 Mt and 20.45 Mt were landfilled and recovered (recycled and energy recovered), respectively [2]. These figures show a substantial potential for C&D waste PwRC production and justify the selection of these states for further investigation.

Recruitment was undertaken according to the Australian National Statement on Ethical Conduct in Human Research [255] and university ethical committee requirements at *RMIT* and *Griffith* Universities. The project's industry partners, *Australia Sustainable Built Environment National Research Centre (SBEnrc)* and the *WMRR*, assisted with the recruitment process by providing their network contact details. *WMRR*'s members consist of businesses and experts who are engaged in recycling and waste management activities. SBEnrc members include government, industry and academic experts involved in built environment issues, most notably C&D waste management. WMRR is the primary industry stakeholder for this work, while *SBEnrc* funded the research project.

Email communication was used as the method of recruitment. In the first round, an email with the project information sheet was sent to a list of participants compiled by the research team. This covered 60 individuals. Since the study required experts in the field, a key selection criterion was adequate experience in dealing with waste management in Australia, so individuals with relevant experience in the waste management and the resource recovery sector were targeted. The list consisted of the two organisation members and other experts identified by the researchers. A reminder email was also sent to those who did not respond in the first round. Interview participation was voluntary, and attending the interview implied informed consent. The investigators maintained the privacy and confidentiality of all interview information as per the human ethics requirements. The interviews were conducted online, using the Microsoft Teams application. Each interview took, on average, 45 minutes. The interview consisted of several questions covering the main issues and opportunities regarding the development of the market for the C&D waste PwRC. As presented in Table 49, the questions selected and analysed for this study sought participant experiences, their opinion on the impact of COVID-19 on C&D waste recovery activities and their perceptions of the issues around the development of the market for C&D waste PwRC.

Table 49. The selected interview questions used in this study

Theme	Questions
Experience	• Could you please introduce yourself and describe your role in your organisation? How long have you been involved in C&D waste management?
COVID-19 and waste management	• In terms of the development of a market for C&D waste PwRC, please share your views on the conditions after COVID-19. • How has COVID-19 affected your organisation/industry? • How does the industry respond to COVID-19 during and after the outbreak?

5.2.2. Data Analysis

Audio data was carefully transcribed by a professional transcriber word-for-word. The research team further verified the quality of text data. Analysis of the transcripts was performed using the NVivo Pro 12 application [256]. This application facilitates codifying text-based qualitative data. A thematic analysis method was applied to identify emerging themes [257], with a combination of inductive and deductive reasoning approaches employed. The analysis began with a deductive or theory-driven coding system (a-priori codes) using the literature review/interview guide elements while creating additional new nodes (in-vivo codes) inductively from emerging interview data.

5.3. Results

The findings are presented under three themes: (1) participant profile including their experience and expertise; (2) COVID-19 impacts on businesses and the C&D waste industry and the construction industry; and (3) responses to COVID-19 impacts and lessons learned.

5.3.1. Participant Profile

The primary stakeholders who have a substantial role in the utilisation of PwRC (e.g., government, recyclers and construction professionals) are well represented in the sample size. As shown in Figure 36, the largest category type was government officials (9), followed by recyclers (8), professionals working in the construction industry (5), experts employed in the construction materials manufacturing sector (3) and consultants with relevant experience (2).

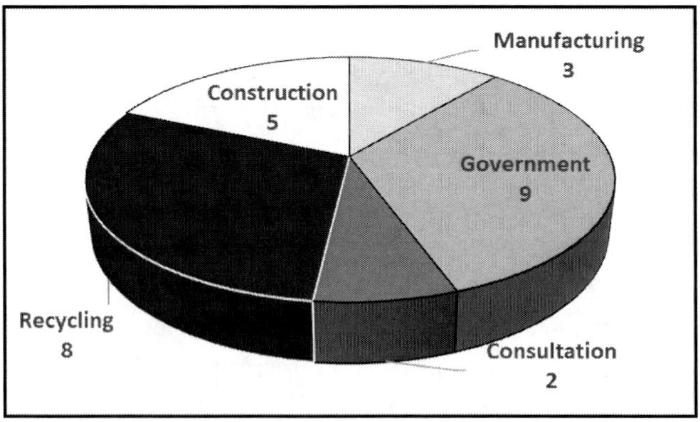

Figure 36. The composition of stakeholder groups among the participants.

The business locations of the participants were WA (n=11), Qld (n=7), Vic (n=5), NSW (n=3) and SA (n=1). This shows a wide geographical spread across Australia. In terms of gender, five participants were female. The remainder were male, which echoes the gender balance in the construction industry. According to the ABS report for 2019-20, the construction industry had the highest proportion of men aged 20-74 at 87% [258].

5.3.2. COVID-19 Impacts on Businesses and the C&D Waste Industry

The COVID-19 pandemic has disrupted the C&D waste industry through changes to working shifts, supply chain interruptions, delays between the C&D cycles, and the inability to retain or recruit a workforce (Table 50). In this study, interview participants from different industry sectors and government shared their lived experiences in the face these disruptions. The impacts varied across AEC sectors; while public construction projects continued, there were significant impacts on residential and commercial projects. Despite the government economic support packages (e.g., JobKeeper [259]), many organisations had to reduce employees on some projects. Furthermore, the number of tenders for construction projects was also reduced. An organisation's ability to continue operating on public projects to a large extent determined their ability to endure the pandemic situation.

Many of the interview participants reported the rapid implementation of COVID-19 safety protocols, including clearly divided shifts, wearing protective equipment and the use of hand sanitiser. Overall, the waste industry addressed many challenges through vertical integration and the continuation of building and construction projects. The following sections describe the key impacts with specific interview examples across the five Australian jurisdictions studied.

5.3.2.1. Interruptions to Regular Business Operations and the Urgent Need for Safety Precautions

Changed work shifts, limited interactions with people, adjusted logistical arrangements, additional safety precautions and minimised onsite engagements impacted regular business operations. There was a significant emphasis on rapid interactions and tasks such as prestart checks and cleaning activities. A private sector representative from NSW shared, *"COVID affected the interactions with the people. Initially, there was a rush on the masks and the hand sanitisers, but as with all free-market economies, a number of companies re-tasked their production and six weeks later none of us had a problem"* [P2]. Given construction's heavy reliance on onsite and in-person work, COVID-19 lockdowns and social distancing measures put a great strain on employees within the industry. For example:

> "We had a few issues around cultural heritage experts coming up from New South Wales to do cultural heritage surveys. So, certain things become challenging that you can't do virtually. When you're going to do cultural heritage surveys, there are certain experts that are required" [P5].

A government representative [P8] from Qld pointed out significant concerns around transport and sanitation. For transport, this related to unknown issues (e.g., tissues, masks) at the time. Government was unable to mobilise and provide clear information immediately. While there was some resource limitation at the beginning (i.e., sanitisers), this was rapidly addressed with hand sanitiser quickly manufactured for industrial use. Meanwhile, paused operations in China impacted the resource recovery sector, and there were concerns about having to stockpile materials unable to be processed or get shipping containers to store the materials.

5.3.2.2. Supply Chain Interruptions

Most participants experienced some form of supply chain interruptions, especially for material purchased from overseas markets. One government participant shared that this was further exacerbated by the quantum of infrastructure projects that were happening simultaneously. For example, they stated:

> "It's also the quantum of infrastructure projects that are happening at the same time, so if you think about, you've got Melbourne Metro, Sydney Metro, Cross River Rail (inland rail). There's a lot of major infrastructure projects that are going to be hitting construction around the same time, and that's going to have an impact only on materials" [P5].

Delays between the C&D cycles, delivering infrastructure projects across the state and delays between the C&D cycles were highlighted as critical impacts. These impacts created higher-order challenges to the organisation in terms of managing the mass balance of C&D required for the sustainable operation of the recycling facility. A participant from SA [P11] confirmed that these delays impacted the state's logistical and resource recovery operations.

Table 50. Summary of COVID-19 impacts on businesses and the C&D waste industry

State	Key impacts
NSW	• A slight decline in demolition type activities from C&D materials received by the site • Decline of waste streams depending on the severity of the state government's reaction • Review of supplier contracts and rates to align with the new state of the economy
Qld	• Supply chain interruptions for material purchased from overseas markets • Limited material flow and increased effects by the quantum of infrastructure projects that were happening at the same time • Interruptions to onsite inspections, especially to cultural heritage surveys • Working remotely and travel restrictions • Restricted production and inability to export and import as usual • Increased levels of efficiency in transportation in certain operations due to clear roads and transport networks • Resource recovery sector impacted by stockpile materials unable to be processed or to get shipping containers to store the materials
Vic	• Increasing budgetary constraints • Dealing with uncooperative industry partners with changed priorities • Decline in market activities in the AEC industry • Delays in several projects starting that is considered an economic slump • Pandemic induced human resource challenges (reduce the number of employees) • Reduced tenders for construction projects
WA	• The waste operation continued as it is considered an essential service • Self-isolation among the workforce in the operations • Incentives in the residential market and not commercial have been counteractive to this impact • Limited resources, both human resources and shortage of materials, particularly steel, timber • Reduced the number of people working on construction sites, particularly building houses • Increased pressure to do better, to recycle more C&D material with increasing infrastructure projects • The residential housing market was doing well, and there has been a positive impact in the broader macroeconomic perspective • Changes in the priority of the organisation activities such as bringing forward of infrastructure projects
SA	• The JobKeeper/JobSeeker schemes negatively impacted labourers' choosing not to do hard physical work while being paid • Increase in workload and the labour hour/cost percentage

5.3.2.3. Pandemic Induced Human Resource Challenges

While many participants expressed significant operational difficulties, contractors—especially ones with large workforces—encountered several pandemic-induced human resource challenges. These include pay cuts, laying off employees or increasing the workload of existing employees. Three participants indicated that their response was to retain the workforce [P5, P6, P11]. One organisation implementing a 30% pay cut across their business to achieve this goal rather than laying off employees. It was considered a fair approach and enabled them to respond quickly to the impacts. For example, as the relevant participant reported:

> "I think everyone felt quite together; everyone felt quite bonded by it because everyone was doing it together. And the people that performed more, the people that were in the higher income of the stream gave up more…as an executive team, we didn't have to start picking winners and losers. And I think that was incredibly motivating for our team, and I think it was incredibly motivating from our managers that, here is a company saying we want all of us to be in this together" [P11].

Further, in this participant's experience, JobKeeper and JobSeeker schemes negatively impacted the hiring of labourers, as many opted to take the government payments rather than doing hard physical work. This phenomenon resulted in a significant increase in workload and the labour hour/cost percentage. Table 51 synthesises the key impacts shared by interview participants across the five states studied.

Table 51. Specific responses to the COVID-19 impacts across five Australian jurisdictions

State	Key responses
NSW	• Managing workforces with clearly defined shifts and cleaning arrangements • Rapid interactions and tasks focused on prestart checks and cleaning • Procuring masks and hand sanitisers • Putting contingency and response plans in place very early and had communication out to the sites • Creating clear communication channels and formulating clear messages
QLD	• Increased attempts to use local products, diversify the supply chain as much as possible to reduces supply chain risks • Leveraging virtual platforms to engage with other colleagues • Creating a circular economy system via local markets and bringing the waste closer and the end products closer

State	Key responses
	• Requesting for possible government funding for the industry to become profitable and to establish • Conducting market assessments to assess the feasibility and to establish in the market • Implementing COVID safety protocols • Focusing on safe ways to work around waste, deliver essential services in local areas and across borders • Creating a group of representatives from all of the impacted components of the waste sector together to discuss issues and create collective solutions • Identifying opportunities related to online platforms and real-time data and digital engineering is positive repercussions of this disruption
VIC	• At an individual level, measures such as social distancing, personal hygiene, taking care of one another, reporting if there is ever an issue with people have symptoms • Repeating the message of the required practices and measures due to C&D waste recycling employees' unique socio-economic background • Placing controls around shared spaces, and customer interaction • Allocation of additional staff to each project called COVID monitors
WA	• Developed targeted protocols in response to COVID-19 situations • The extra construction work that has been undertaken is used up to stock balls or had some of those • raw materials being turned into, such as recycled blocks or recycled road-based • Creating a memorandum of understanding to share sources depending on the situation • Release of planning caveats around the COVID-19 period to stimulate the economy • Application of the segregation and the work-at-home requirements advised during the lockdowns • Followed the advice from the state health authorities such as social distancing, putting a face mask • The government's response was to try and fast track infrastructure projects as part of the economic recovery, which was an excellent opportunity for the government to increase the use of recycled C&D • Working from home and preparing for the booming residential market • The government responded to the COVID-19 situation by supporting the use of recycled content and increase in the number of projects that which has broadened the opportunities
SA	• The industry and government worked together to address COVID-related issues • The organisation implemented a 30% pay cut across their businesses to achieve this goal instead of laying off employees

5.3.3. Responses to COVID-19 Impacts and Lessons Learned

Participants across the five Australian jurisdictions shared their experiences relating to the responses to COVID-19 disruptions and challenges. While most participants described their rapid responses as aligned with emergent safety priorities and efficient management of resources, the experience also helped them to uncover a range of opportunities. In adapting to the realities of the COVID era, participants also developed consciousness around the need to leverage digital technologies, devised business contingency plans, created coalitions between government and industry, and diversified supply chains to reduce supply chain risks. The key responses are as follows.

5.3.3.1. Leveraging Digital Technologies

The emergent opportunities using real-time data, digital engineering and other online technologies when dealing with such disruptions were highlighted by several participants. For example, P6 stated, *"I think COVID might have positive repercussions for things like smart cities and real-time data, and digital engineering. I think that those shifts would then support this kind of work."* Another participant engaged in a large transport infrastructure project reiterated the importance of leveraging online technologies [P10]. One government sector participant shared their experience leveraging current digital capabilities of using cloud-based technologies to work with clients and colleagues across different states, explaining, *"I think we'd already moved to before COVID because we work across three different states. And GIS has been used for modelling, I think we were set up quite well, virtually, prior to COVID, simply because of our expansive breadth"* [P5].

5.3.3.2. Developing Contingency Plans

The importance of contingency plans within the business continuity planning process was emphasised by two participants. For example, one participant shared their experience on developing contingency plans:

> "Our leadership team identified the potential impacts that COVID was going to have. We put contingency plans in place, response plans in place very early and had communication out to the sites, the ability to work remotely was responded to quite quickly. So as a business, and in terms of potential impacts, we had really good continuity plans in place" [P1].

5.3.3.3. Creating Coalitions between Government and Industry

There was evidence of efforts to collectively create a coalition between government and industry to address the emergent issues around the pandemic collectively. This group consisted of representatives from all the impacted components of the waste sector together to discuss issues and create collective solutions. These issues captured, *"Will there be enough capacity to deal with waste? how to treat mixed waste containing medical waste? What were the risks to landfill operators?"* [P9]. There was also evidence from SA where organisations have engaged strongly with the government, the industry, and their customers. This has resulted in the organisation's preparedness and to be more resilient in the market. However, two government representatives [P24, P27] mentioned that due to the government COVID-19 recovery plan in response to the COVID-19 situation, which includes bringing the infrastructure projects forward to support local industry, the market would be more flourishing for C&D waste streams after COVID-19. At the industry level, several participants stated that during COVID-19, the industry and government worked well together in response to issues that come with the pandemic. One interviewee had participated in several industry leadership calls with various state public organisations to communicate issues the industry face during the affected period. The industry itself was quick in taking preventative containment measures. For example, *"that's what I think was interesting, I mean you look at it and you thought, and this was in normal circumstances, the entire industry would say they need three years to implement these types of changes and they happened in three weeks. So, for me that was a really interesting phenomenon that we saw right across the industry"* [P11].

5.3.3.4. Diversifying Supply Chains and Reducing Supply Chain Risks

Diversifying the supply chain and creating more local transparent supply chains were suggested as key measures to mitigate risks. One participant stated that as their business operations are located in China, this provided them with some early opportunities to learn and navigate through disruptions, and they could be ahead of the curve [P6]. There were early efforts to set up team-based systems, understand their procurement needs and mitigate the risks. Another participant reiterated the importance of reducing risks through creating secondary markets for C&D waste and enabling local supply chains. For example, they contended, *"I think any country should be looking at ways to reduce their exposure to supply chain risk, and if you can keep things onshore, that does a variety of things. It reduces supply chain risks, it creates jobs, and it creates alternative industry"* [P5].

5.3.4. Emergent Impact Matrix

As outlined in the Results section, the impacts of the COVID-19 pandemic on the industry in the studied states have been diverse. Table 52 provides an impact matrix developed based on the participant responses. The impacts are categorised under five scopes: 'construction and waste generation', 'employee engagement', 'transportation', 'processing and recycling' and 'disposal'.

Table 52. Summary of COVID-19 impact matrix in the five studied states

Scope	Impact description	States				
		NSW	QLD	VIC	SA	WA
Construction and waste generation	Interruptions to regular business operations	●	●	●		●
	Urgent implementation of safety protocols	●	●	●	●	●
	Decline of waste streams/limited material flows	●	●			●
	Supply chain interruption	●		●		
	Decline in demolition type activities	●				●
	Economic impacts/budget constraints					●
	Changing of priorities			●		●
	Increased infrastructure development due to government interjections			●		●
	Reduced tenders for construction			●		
	Delays in project planning and delivery			●		●
Employee engagement	Immediate need for remote working arrangements		●			
	Limited onsite engagement		●			
	Pandemic induced human resource challenges (layoffs, pay cuts, increased workloads)			●	●	
Transportation	Restricted travel		●			
	Waste transportation and handling risks					
	Increased efficiency due to clear roads	●	●			
Processing and recycling	Restricted production					
	Decline in demolition type activities/limited material flow					
Disposal	Stockpiling of materials unable to process		●			

Table 52 shows that the category participants mentioned most was the construction and waste generation impact category, followed by employee engagement. Furthermore, only 'urgent implementation of safety protocols' was reported as an impact across all study states. Interestingly, participants did not report restricted production and decline in demolition activities occurring during pandemic restrictions. This is mainly related to state-level COVID-19 policies, which consider construction and recycling to be essential industries.

5.3.5. Adaptive Measures

Based on the findings, the following is a summary of adaptive measures relevant to the C&D waste industry that were undertaken at a governmental level in response to the pandemic.

5.3.5.1. Government Information Support

State and territory governments have dedicated information hotline numbers for businesses affected by COVID-19. These information centres operated in the ACT, NSW, Vic, Qld, WA and the NT. Notably, in Qld, the Chamber of Commerce & Industry Queensland has provided dedicated advisory services for small businesses. The federal government's business hotline also provides support for impacted small and medium enterprises.

5.3.5.2. Government Financial Support

State and territory governments announced financial support for the businesses affected by the pandemic (More information on this support can be found via the links in Table 1). As expected, the highest level of financial supports was offered by the Qld ($2.5 bn), NSW ($2.3 bn) and Vic ($1.7 bn) governments due to the comparatively larger size of their economies and populations. Qld and ATC governments provided electricity rebates with a specific consumption threshold. All jurisdictions offered payroll tax relief in some capacity [260], between 3–12 months in duration. The federal government financial support was applied through three schemes: the Economic Recovery Package (ERP), JobKeeper Payment scheme [255], and Tax-Free Cash Payment (TFCP). ERP provided support for the Australian economy, including small to medium enterprises, for three years. JobKeeper aimed to support employers to keep their staff; the eligible business had to prove that their turnover was affected by COVID-19 by a decline of between 15–50%. TFCP provided tax-free payments between $20–100k. The government committed more than $4.7 m to provide small regional businesses affected by COVID-19 with access to free and confidential financial counselling. In NSW and Qld, affected small businesses could apply for a Re-opening grant (a one-off $3,000) and COVID-19 Adaption grants ($10k), respectively.

In Australia, except for a few industry-led inquiries, research on the COVID-19 impacts on waste recovery is limited. One example is the survey facilitated by the ACOR to further understand stakeholders experience during the pandemic. The majority of responses to the ACOR COVID-19 Industry Pulse Check survey suggested a landfill levy relief for the next 6–12 months [247]. Other proposals included local governments waiving business fees; state and local governments becoming more flexible on certain facility license conditions for social distancing so the protection of staff could be maintained; a landfill levy waiver on contaminated residuals from recycling facilities; payroll tax reductions; deferring business loan repayments; lifting collection time curfews so that bins could continue to be collected; and further support for tax and super obligations.

5.3.5.3. Interstate Waste Movement

The primarily effects on interstate waste movement occurred due to border restrictions imposed by Australian jurisdictions in an effort to stop the spread of COVID-19. NSW, Qld, Tas, SA, WA and the NT imposed movement restrictions across borders, but there were exemptions for essential travel, workforce movements, freight and specific compassionate grounds. For instance, NSW Public Health (COVID-19 Border Control) Order 2020 permitted critical services, including construction and the maintenance and repair of critical infrastructure. The *NSW EPA* information fact sheet (April 2020) specified that the authority was not imposing restrictions on interstate transport of waste [261]. In the NT and Tas, interstate travellers had to complete 14 days of self-quarantine unless they met exemption criteria. Travellers from declared hotspots to Qld were also required to complete 14 days' quarantine in government arranged accommodation (at their own cost) unless they were exempted from the requirements. In SA, while the board restrictions eventually ceased for travellers from WA, Tas and the NT, travellers from other states were subject to an assessment of the public health risk. In WA, restrictions were put in place that disallowed people entering the state without an exemption.

Conclusion

This chapter provides useful insights regarding the impact of the COVID-19 pandemic on the C&D waste management and resource recovery industry in Australia. The interviews identified that the pandemic affected industries in various ways (e.g., work disruption, extra H&S implications, economic and environmental performance). As such, there were some specific impacts on the C&D waste industry (e.g., continued drive for homebuilding, supply chain issues). With regards to recycling, impacts were mixed. In some instances, the sector reverted back to unsustainable landfill and incineration; in other instances, the situation inadvertently encouraged increased recycling practices.

This study suggests the following policy recommendations to further assist the industry with sustainable operation during challenging times:

1) Consideration of the industry as an essential service in all Australian jurisdictions.
2) Landfill levy relief for the next 6–12 months to stop illegal dumping and stockpiling.
3) A landfill levy waiver on (contaminated) recycling residuals.
4) Collection time curfews to be lifted so that bins can continue to be collected.
5) In response to disruptions, such as from COVID-19, the C&D industry needs to embrace digital and technological developments in order to survive and thrive.

While the study capitalises on the participants' knowledge and extensive industry experience, there is a limitation concerning the timing of the interviews (conducted in 2020) in terms of the information gleaned in relation to a fast-moving pandemic. At the time of publication, the situation has evolved, though the full impacts of the COVID-19 pandemic are still yet to be known. The findings from this study may be superseded by more important outcomes later on. However, these interviews did generate some interesting insights and lessons for other contexts. Hence, it is proposed that further studies investigate how the proposed adaptive measures from this study, and others subsequently emerging, could be practically implemented and enable a resilient C&D waste management and resource recovery industry.

Chapter 6

Optimal Model for the Creation and Stimulation of End-Markets for C&D Waste Streams

6.1. Introduction

Australia's capital cities are experiencing unprecedented construction activity to meet the needs of the growing population and keep up with the economic growth, resulting in the issue of managing the ensuing C&D waste coming to the forefront [262]. The use of PwRC has been highlighted by the main stakeholders, as explored in Chapter 1, as a key solution to effectively manage the substantial quantity of C&D waste. Yet, as Chapter 3 showed, currently, there is a limited market for C&D waste PwRC in Australia. Consequently, in this chapter, efforts are made to identify strategies that can stimulate the end-market for this waste stream, culminating in the proposal of a model designed for this purpose.

6.2. Approach

This chapter builds on the primary data collected through interviews with experts who have relevant experience in managing C&D waste in Australia, as well as the background and context explored in the preceding chapters. The content of this chapter is based on the similar approach that is used and described in Chapter 5 (refer to Section 5.2 for further information). The interview questions that were used during data collection are presented in Table 53.

Table 53. Research questions used in this study

Theme	Questions
Experience	• Could you please introduce yourself and describe your role in your organisation? Prompt: involvement/experience in C&D waste management • How long have you been involved in C&D waste management?
C&D waste PwRC market development	• Do you think that creating an end-market for C&D waste could be used as an intervention to manage C&D waste better? • What is your opinion about the operation of existing end-markets in your jurisdiction? • What can be improved to facilitate creating and stimulating efficient end-markets for C&D waste in your industry/state? • Which C&D waste materials have the most potential for new markets? Prompt: concrete, brick, steel, glass, timber, asphalt and clean fill • In your opinion, how can we enable the creation of an end-market for C&D waste? • What are the main factors impacting C&D waste end-of-life management? How are they influencing C&D waste management according to the status quo? • What are the main barriers to implementing a sustainable end-market for C&D waste in your industry/jurisdiction? • How do you think of the current legislation about C&D waste end-of-life management in your jurisdiction?

6.2.1. Participant Profiles

A summary of participant's profiles, including the primary location of their business and the focus of the operation, and their position and length of experience, is provided in Table 54. The locations of the participants' businesses were in WA (n=11), Qld (n=7), Vic (n=4), NSW (n=3) and SA (n=1). In terms of gender, five participants were female, and the rest were male, echoing the industry gender imbalance.

6.3. Results

6.3.1. NSW

The participants based in NSW were representatives of two private organisations, including one recycling facility and one manufacturing company (Table 54). Three of these research participants held senior leadership and managerial positions at the time of the interview.

Table 54. Summary of profiles of participants and their organisations

P	State	Stakeholder group	Position	Experience (years)	Recent involvement in C&D waste management
P_1	NSW	Manufacturing	State Manager	4	Providing the construction industry with sustainable products and solutions in NSW and Vic
P_2	NSW	Recycling	Director	30	Producing and re-selling quarried and recycled landscape products in NSW; waste management, crushing and landfilling operations, technical management of C&D waste resources in NSW
P_3	NSW	Manufacturing	Strategy Business Development Manager	4	Providing the construction industry with sustainable products and solutions, and business development related to C&D recycling
P_4	QLD	Government	Sustainability Manager	20	Managing C&D waste associated with large infrastructure projects
P_5	QLD	Government	Sustainability Manager	11	Managing C&D waste associated with large infrastructure projects
P_6	QLD	Construction	Sustainability Operations Manager	20	Designing, delivering and reporting on activities as a business for sustainability metrics
P_7	QLD	Construction	Senior Waste and Resource Consultant	13	Investigating the regulations and legislation in different jurisdictions, the impacts of establishing reprocessing facilities and conducting market assessments
P_8	QLD	Government	Environmental Coordinator for transport and Infrastructure	14	Managing statutory approvals for all applicable infrastructure and maintenance activities, environmental matters and contract support for other parts of local council
P_9	QLD	Government	Executive Director of Resource Recovery	16	Developing long-term policy levers that can be used to drive positive change and shift behaviour and language from waste to resource recovery
P_{10}	QLD	Consultation	Managing Consultant	10	Providing expert advice on managing different streams C&D waste across infrastructure

Table 54. (Continued)

P	State	Stakeholder group	Position	Experience (years)	Recent involvement in C&D waste management
P_{11}	SA	Recycling	Chief Executive Officer	10	Resource recovery of C&D waste materials and tyre recycling in SA; sells PwRC in various markets
P_{12}	Vic	Recycling	Chief Executive Officer	18	Recycling C&D waste materials in Vic; has the largest C&D waste recycling facility in Australia
P_{13}	Vic	Government	Environmental Project Advisor	4	R&D and market development for PwRC, resource recovery and facilitating the connection between research institutes and the industry to produce PwRC in Vic
P_{14}	Vic	Recycling	Sales Manager	8	Liquid waste management, managing crushing and landfilling operations, and technical management of C&D waste resources recovery activities in NSW and Vic
P_{15}	Vic	Recycling	Managing Director	25	Managing recycling of C&D waste, particularly asphalt, in Melbourne and Brisbane
P_{16}	Vic	Construction	Quality and Environment Manager	18	Monitoring C&D waste disposal and use of PwRC, and improving recycling waste generated at construction sites in Vic, SA and some contribution to other areas
P_{17}	WA	Consultation	Principal Engineer Pavements & Drainage	13	Providing consultation services pertaining to sustainability in the construction industry and lecturing in the University of Western Australia about C&D waste management
P_{18}	WA	Recycling	General Manager	12	Recycling C&D waste materials in WA
P_{19}	WA	Government	Sustainability Manager	5	Large-scale demolition projects across WA with the aim of recycling C&D waste
P_{20}	WA	Government	Senior Development Manager	14	Engaging in residential development sites including demolition of old buildings, remediation of sites and civil bulk earthworks for development
P_{21}	WA	Recycling	Resource Development Manager	10	Recycling and managing C&D waste in WA, and also representing the recycling industry in different state and national committees

P	State	Stakeholder group	Position	Experience (years)	Recent involvement in C&D waste management
P_{22}	WA	Recycling	Director-operations	5	Providing technical solutions to manage the traceability of materials during deconstruction projects
P_{23}	WA	Manufacturing	Technical Manager	15	Managing mix designs in production line and performing troubleshooting and trials relating to quality control for concrete
P_{24}	WA	Government	Manager – Policy	16	Delivering the state waste strategy in two public organisations
P_{25}	WA	Construction	Residential Sustainability Manager	4	Improving sustainability aspects of residential construction projects
P_{26}	WA	Government	Manager of Material Engineering	19	Contributing to planning for creating road infrastructures in WA in a public organisation with some involvement in using PwRC in these projects
P_{27}	WA	Government	Principal Advisor Sustainability	15	Responsible for corporate sustainability in two public organisations, which involves waste management and market development for C&D waste materials

The participants' experience related to the C&D waste management field ranged between 4 [P_1] and 30 years [P_2]. The main focus of participants consists of producing and re-selling quarried and recycled landscape products in NSW [P_2]; waste management, crushing and landfilling operations, technical management of C&D waste resources in NSW [P_2]; and providing the construction industry with sustainable products and solutions [P_1, P_3].

6.3.1.1. The Impact of COVID-19 and Responses

The COVID-19 pandemic has created challenges for various sectors, including housing, building and infrastructure. Many businesses have experienced some form of disruption through changes to their working shifts or supply chain interruption. The interview participants shared their lived experiences in the face of these COVID-19 disruptions. While many of them acknowledged the rapid implementation of safety protocols, including changes to shifts, wearing protective requirements and use of hand sanitisers, the overall waste industry processes were insulated because of vertical integration and the continuation of building and construction projects. There was a significant emphasis on rapid interactions and tasks such as prestart checks and cleaning activities. For example, a private sector representative shared:

> "COVID doesn't affect the basic process, it's only the interaction with the people. Initially, there was a rush on the masks and the hand sanitisers, but as with all free-market economies, a number of companies re-tasked their production and six weeks later none of us had a problem" [P_2].

Another private participant highlighted the importance of contingency plans business continuity planning to keep the business running. For example:

> "Our leadership team identified the potential impacts that COVID was going to have. We put contingency plans in place, response plans in place very early and had communication out to the sites, the ability to work remotely was responded to quite quickly. So as a business, and in terms of potential impacts, we had really good continuity plans in place" [P_1].

The detailed impact and responses to COVID-19 are tabulated in Table 55.

Table 55. The impact of COVID-19 and the responses to the impact in NSW

P	Impact(s)	Responses
P_1	While work was not directly impacted, a slight decline was witnessed in demolition type activities from C&D materials received by the site. From an outbound sales point of view, there was minimal impact on the construction market around lockdown, but they managed to continue with sales.	From an operating cost point of view, a decision was made around managing workforces to ensure particular day shift and afternoon shift teams that didn't cross over. There was always a gap between a day shift and an afternoon shift to clean the common facilities. These cleaning procedures increased costs slightly, around labour costs particularly, and the cleaning costs associated with keeping the businesses as free from direct impact from positive cases as possible.
P_2	It was evident that some waste streams slowed down in various capitals, depending on the severity of the state government's reaction.	There were rapid interactions and tasks focussed on prestart checks and cleaning. While COVID did not affect the basic process, it only impacted the interaction with the people. Initially, there was a rush on procuring masks and hand sanitisers, but as with all free-market economies, several companies re-tasked their production, and six weeks later, none of them had a problem.
P_3	There was a normal impact throughout the economy. Some efforts were in place to push for the procurement team to reach out to all suppliers trying to review contracts and rates to align with the new state of the economy. It was considered that construction industry was the least impacted by COVID, as it was positioned as an essential activity.	The leadership team identified the potential impacts that COVID was going to have and made early efforts to put contingency and response plans in place very early and had communication out to the sites, so the ability to work remotely was responded to quite quickly. That meant, as a business, and in terms of potential impacts, they had excellent continuity plans in place.

6.3.1.2. Market Development in NSW

There was consensus among the participants on the benefits of developing the market for better management of C&D waste materials in NSW (Table 56). Factors such as location, recycling rates, waste levy, incentives and products specification were highlighted as key characteristics to consider when creating end-markets for C&D waste. For example:

> "And the other element that comes to mind is the incentives on the supply and the demand side of things. So, for the supply, things such as landfill levy fees, we'd put some incentives for the supply of those involved in the recycling operations" [P_3].

Table 56. An overview of perceptions of market development for C&D waste PwRC in NSW

P	Intervention	Operation	Potential markets
P_1	The resource recovery rate for C&D materials is quite high in NSW, and mature end-markets drive it for the material.	The end-market use of the product is quite mature in NSW, but there is the capacity to improve the market demand through improved product specifications and incentives.	In the industry, there is more focus on source-separated concrete brick and asphalt. In terms of exploring other waste streams, it is critical to have methods of separating those resources early on so that they are not co-mingled to allow for better recycling of the materials in their purest form.
P_2	There are intentions to move to C&D sourced recycled products and it is important clearly understand natural and recycled material standards.	There is a scaling system now, which gives higher scores towards recycled products, so recycled is sought when contractors bid for big infrastructure projects, but it does not have enough weight. Two recycled products are not allowed to be mixed in NSW. This means that suppliers can safely deliver recycled soil and recycled mulch to a site and mix it.	Nearly 30% of C&D waste is timber in NSW, and there is more potential to re-use and recycle this material.
P_3	There are several factors affecting this intervention: geographic location, recycling rates, product specifications and relevant opportunities in infrastructure projects (i.e., road-based projects) to use PwRC.	There are regulatory requirements around minimum standards now for C&D facilities. However, within the organisation, there are rigorous inspection and receivable processes, which start with inbound account managers going out to sites where the demolition or the excavation works are being undertaken. Regularly, they check clearance certificates where necessary, where there has been the risk of asbestos associated with the work.	It's very important to create a cleaner waste stream. The more waste streams are separated, the more opportunity there will be to recycle them into a higher value product. Plastics and metals are new opportunities for market development.

They also expressed their frustrations about the current market regulations and product specifications as impediments to creating end-markets C&D waste. The participant compared NSW with other states and expressed their frustration about the current EPA system. For example:

> "The other states are bit more circumspect and pragmatic, but New South Wales has a serious problem, and the main problem is the concept of asbestos contamination" [P_2].
>
> "This participant also pointed out that the current market criteria and specifications are hindering the recycled product market. For example, "The C&D products are intended to replace natural products, but unfortunately in many cases, the state-based intra-mentalities will apply exactly the same natural material criteria as to recycled material" [P_2].

In terms of the potential market, responses from two participants [P_1, P_3] underlined the potential market for plastics and timber. Notably, one participant emphasised the importance of having a clean stream of waste to create a value-added product, stating, *"The more you can keep them separated, the more opportunity you've got to resource or recycle them up back into a higher value product"* [P_3].

6.3.1.3. Improvement Areas

The industrial practitioners shared a range of insights linked to improvement opportunities in relation to the creation and simulation of efficient end-markets in NSW. One participant emphasised that the government needs to mandate a certain percentage of recycled content on all government jobs and thereby advance the recycling practices. For example:

> "You have to make it financially worthwhile to recycle, but you also have to create the regulations such that the markets will be there. Again, you see that not occurring in our state, whereas if they mandated a percentage of recycled content in a finished product on all government jobs, for example, preferential content, that would really further drive recycling" [P_2].

Going forward, one participant contended, business needs to be encouraged to shift from merely being compliant to taking additional efforts to engage in sustainable business practice, that is, moving away from *"That compliancy focus for the recycled operations and finding the right balance there between establishing strict parameters for the recycling operations to operate under, versus not making those operations through costing"* [P_3].

A more streamlined, harmonious approach to approvals and timeliness of the approvals, and early engagement and support from the regulators, were identified as critical to improving end-markets for C&D waste. These factors will also influence the investment decisions of market players. For example, one participant said, *"Because we'll make the decision that it's worth making the investment as opposed to saying that money is better elsewhere"* [P_1].

Furthermore, considering the circular nature of the economic system and tracing the material movement was also identified as a critical area of improvement. One participant explained:

> "The whole CE element that we need to think in the 20-year Waste Strategy, that we are going to start seeing more circulatory thinking, which sees materials coming in and out of C&I and C&D, moving through a broader lifecycle process for materials" [P_1].

6.3.1.4. Enablers of and Barriers to Market Development in NSW

This section presents key enablers and barriers described by the interview participants. As summarised in Table 57, there are myriad enablers and barriers related to the market development in NSW. The main enablers in NSW were supportive and straightforward policies and specifications augmented with a streamlined, harmonious approach to approvals, timeliness of the approvals, and early engagement and support from the regulators. In terms of barriers, all responses pointed to the risk of contamination (i.e., asbestos), strict product specifications, lack of data traceability of material movement and irresponsible behaviour of rogue operators. For example, one participant described the impact of the irresponsible behaviour of rogue operators: *"It's those that don't follow that strict process, that allows contaminated recycled material to go into the market which then creates the nervousness in the market"* [P_1].

One participant notes that a better understanding of the material movement will provide more opportunities for industries to make their investment decisions because, *"We can then decide where should we continue to invest our time and infrastructure in order to get the most competitive products back into the market"* [P_3].

Table 57. Summary of enablers and barriers specified by the participants in NSW

P	Enablers	Barriers
P_1	• Having specifications that facilitate the recycled content • Shift from the compliancy focus for the recycled operations and finding the right balance when establishing strict parameters for the recycling operations • A more streamlined, harmonious approach to approvals, timeliness of the approvals, and early engagement and support from the regulators	• Lack of data regarding the long-term performance of PwRC • Exposure of particularly asbestos contamination within the materials • Irresponsible behaviour of market operators creating nervousness in the market
P_2	• Employing relevant technologies to support recycling • Legislating a preference for recycled products • Increasing awareness of recyclers network and supply chain • Supportive regulations	• The state-based intra-mentalities will apply exactly the same natural material criteria as recycled material • The concept of asbestos contamination • Rapidly anti-business EPA that is run by government bureaucrats
P_3	• Considering more circulatory thinking • Evaluating the material cycles, especially materials coming in and out of C&I and C&D, moving through a broader lifecycle process for materials	• The limitations and specifications for recycled products is limiting the value creation opportunity from a more CE perspective • Keeping those resources at the highest value within their lifecycle • Lack of understanding of how to create CE with construction materials to really drive a lifecycle on those products through design • Lack of information, lack of data and lack of traceability • Lack of understanding of the material movements • Lack of a mature market leads; customers to choose not to use PwRC because of the risk of contamination • The lengthy periods are taken for approvals from government authorities

6.3.1.5. Influential Factors for Materials End-of-Life Management in NSW

The participants discussed several influential factors for materials end-of-life management. These include dealing with many unknown types of material and different consistencies of materials coming in from multiple customers, and the efforts to put through a process and develop a consistent quality product out the other end. The factors specified by participants included a mix of limited incentives and investment in innovation in the recycling sector, social and environmental issues, and difficulties to self-regulate. For example, one participant explained the higher-order impacts on the communities if industrial practitioners do not take responsible steps to manage C&D waste throughout the lifecycle effectively:

> "In C&D recycling, you're dealing with lots of unknowns of different types of material and different consistencies of materials coming in from multiple customers that you're trying to put through a process and develop a consistent quality product out the other end. And if you don't care to put the effort in through that process, you will accept rubbish into your site, and contamination, and therefore you are putting that out into the community, which exposes the community, which is unacceptable but also puts stress on our industry, as to the standards that are then expected and applied are set to the lowest bar" [P_1].

6.3.1.6. Waste Regulations in NSW

The responses to the question related to current regulations and policies revealed a variety of insights (Figure 37). All three participants expressed both positive and negative perceptions of relevant regulations (Table 58). Specifically, P_1 described opportunities for overall improvement of existing legislation, stating:

> "…we see that there's a huge opportunity in continuing the path of what we've done in New South Wales from a national perspective. The hesitation for us comes from the points we've just made around inconsistencies of legislation, inconsistencies of materials in those markets, uncertainty around legislation and the approval processes" [P_1].

These strict regulations lead to uncertainty to invest, discouraging many players in the market from engaging in the PwRC business. For example:

> "The one thing that does concern us greatly, is the risk of change to the legislation, particularly in the orders and exemptions. The EPA fought rightly or wrongly to make changes to a number of orders and exemptions recently in the end-use space that effectively destroyed that market. So that's probably a risk for us in the regulatory environment is some of the uncertainty around the legislation or the orders and exemptions

particularly that sit in the legislative environment to give us that uncertainty to invest" [P_1].

This participant highlighted the significant need for policy reform to allow some level of recycling material trials and waste recovery, explaining:

"The resource recovery orders and exemptions for their intention work well but need reform. And that's just about being clear around how the resource recovery exemptions operate and being flexible enough that as the markets start to move—and we're looking at different ideas and uses for recycled materials—that there's sufficient flexibility in that resource recovery order exemption framework to allow for trials of materials" [P_1].

6.3.2. Qld

The participants based in Qld represented three government organisations, two private consulting companies and two construction companies (Table 54). Four out of seven participants held senior leadership and managerial positions at the time of the interview. The participants' exposure to C&D waste issues ranged between 10 [P_{10}] and 20 years [P_4, P_6]. The main focus of participants were managing C&D waste associated with large infrastructure projects [P_4, P_5]; developing long-term policy levers that can be used to drive positive change [P_9]; designing, delivering and reporting on activities as a business for sustainability metrics [P6]; investigating the regulations and legislation in different jurisdictions and impacts of establishing reprocessing facilities [P_7]; and managing different streams of C&D waste across infrastructure [P_{10}].

6.3.2.1. The Impact of COVID-19 and Responses

Participants from Qld shared their lived experiences related to the impacts of COVID-19 on C&D waste recycling activities. While most participants described their rapid responses aligned with emergent safety priorities and supply disruptions, they also noted that this experience has also helped them to uncover a range of opportunities. Most participants experienced some form of supply chain interruption, especially for material purchased from overseas markets. One government participant shared that this was further exacerbated by the number of infrastructure projects that were happening at the same time.

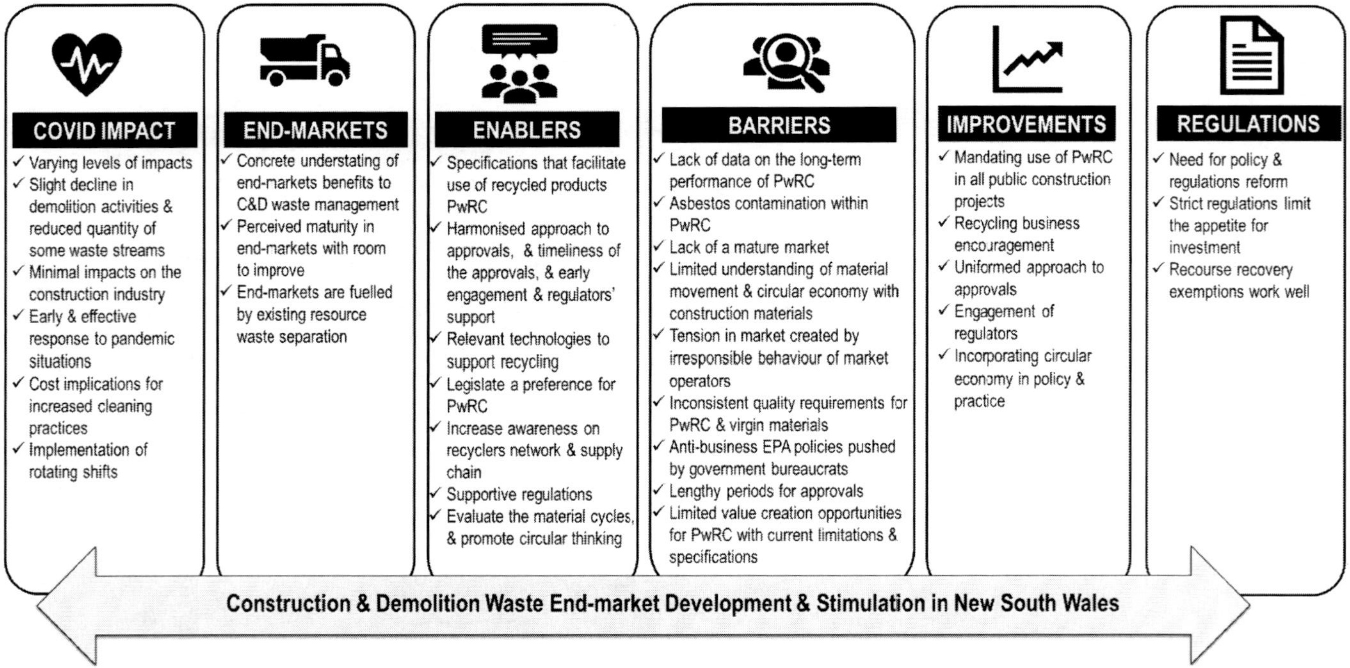

Figure 37. A conceptual model that describes C&D waste end-market characteristics in NSW.

Table 58. Summary of participants' opinions on the impact of current legislation on market development in NSW

P	Policy/Regulations
P_1	It is not an attractive sector because it is so highly regulated, and there are high risks of contamination. To address these risks, it is critical to follow appropriate tests and all those processes.
	While legislation has the right intent, there are some issues with the current EPA, particularly around changes that were made to the POEO Act with asbestos waste reforms. In terms of testing for asbestos, before it arrives at a recycling site, it is a choice of the developer as to whether or not they do the asbestos clearance certificates.
	There are inconsistencies in the legislation around the interpretation of asbestos contamination that need to be resolved. Within this context, dealing with local approvals is a very cumbersome process; for example, it may include approvals from different local and state-level authorities. Getting them all to agree and come together to permit a businesses' intentions in creating those end-markets can be difficult.
P_2	The limited representation of scientists and engineers in the NSW EPA is a challenge for data communication. This government body is an emotional or philosophically-based regulator rather than a science-based regulator. This leads to an uncertain external environment for C&D recycling in the Sydney region. If the existing regulation is enforced, most of the businesses will exit the recycling business because of the significant risk.

Table 59. The impact of COVID-19 and the responses to the impact in Qld

P	Impact(s)	Responses
P_4	There were no remarkable changes in the organisation's functions. A large priority infrastructure project continued with no tangible changes.	There were no tangible changes.
P_5	There were some supply chain interruptions, especially for material purchased from overseas markets. This was exacerbated by the number of infrastructure projects that were happening at the same time. There was some level of interruptions to on-site inspections, especially to cultural heritage surveys.	Increased attempts to use local products as much as possible to reduces supply chain risks. The benefits of local procurement include creating jobs and alternative industries. More functions were carried out through virtual platforms as construction continued; colleagues were predominantly engaged with via online communication platforms such as Skype or Zoom. Furthermore, there were opportunities for professional engagement as there were many infrastructure projects and more suppliers were keen to get involved.

Table 59. (Continued)

P	Impact(s)	Responses
P_6	All projects managed to operate during COVID disruptions. It has not impacted the ability to continue work; however, global supply chains were challenged.	Efforts are being made to try to diversify the supply chain to mitigate risks. Business operations are located in China, so this provided some early opportunities to learn and navigate through disruptions, and be ahead of the curve. There were early efforts to put systems in place for teams, understand their procurement needs and mitigate the risks.
P_7	Experienced working remotely, travel restrictions, restricted production, limited material flow and the impacts on markets not been able to export and import as usual.	The situation was an eye-opener, especially to the potential to create a CE system via local markets and to bring the waste and the end products closer together. There was a suggestion of possible government grants and funding for the industry to become profitable and to establish these systems. More market assessments were taking place in Qld through the firm for industries planning to establish themselves on the market.
P_8	There were minimal impacts. While there was a consideration of whether the construction project should continue, that was managed through targeted risk management strategies. Some increased levels of efficiency were experienced in certain operations due to clear roads and transport networks.	There were COVID protocols for people working on site along with social distancing, with limited visitors.
P_9	• The organisation was not as heavily impacted as many other areas, and most of the concerns for the waste sector were around transport and sanitation, especially transport related to unknowns (e.g., tissues, masks) at the time. They were unable to mobilise and provide clear information. While there was some resource limitation at the beginning (i.e., sanitisers), this was rapidly addressed by manufacturing hand sanitiser for industrial use. • The local government became very concerned for a while about contact with the public, due to people being bored at home and deciding a trip to the transfer station or a landfill was a great day out.	The responses were very responsible with a special focus on safe ways to work around waste and deliver essential services in local areas and across borders. From a government perspective, when COVID first started, they created a group of representatives from all of the impacted components of the waste sector to discuss issues and create collective solutions.

P	Impact(s)	Responses
	The C&D sector has not witnessed significant impacts in terms of volume reductions and so forth because of COVID. As the construction sector has continued to undertake its work, the waste industry has continued its work accordingly as well. The resource recovery sector was already impacted by paused operations in China, but there were concerns about having to stockpile materials unable to be processed or to get shipping containers to store the materials.	
P_{10}	The industry broadly has carried on somewhat unscathed compared to other sectors.	The COVID situation prompted more innovation and things started to be done differently. The shift to online platforms and real-time data and digital engineering are some positive repercussions of this disruption.

Furthermore, on-site engagements, including inspections and surveys, were interrupted. Learning from these COVID realities, participants developed consciousness around the need for business continuity, contingency planning and more transparent, local supply chains.

One private sector participant highlighted the opportunities granted by using real-time data, digital engineering and other online technologies when dealing with such disruptions. For example, P_6 stated, *"I think COVID might have positive repercussions for things like smart cities and real-time data, and digital engineering. I think that those shifts would then support this kind of work."* Another participant engaged in a large transport infrastructure project re-iterated the importance of leveraging online technologies [P_{10}]. One government sector participant shared their experience on leveraging the current digital capabilities of cloud-based technologies to work with clients and colleagues across different states, explaining:

> "I think we'd already moved to Adobe Sign prior to COVID, because we work across three different states. And GIS has been used for modelling, I think we were set up quite well, virtually, prior to COVID, simply because of our expansive breadth" [P_5].

There was also evidence of efforts to create a coalition between government and industry to collectively address the emergent issues around the pandemic situation. This group consisted of representatives from all of the impacted components of the waste sector to discuss issues and create collective solutions. These questions captured, *"Will there be enough capacity to deal with waste? How to treat mixed waste containing medical waste? What were the risks to landfill operators?"* [P_9]. The detailed impact and responses to COVID-19 are tabulated in Table 59.

6.3.2.2. Market Development in Qld

There was general agreement among the interviewed stakeholders on the importance of creating a market for better management of C&D waste materials in Qld (Table 60). Factors such as early engagement of suppliers [P_9], waste levies [P_7], take-back policies [P_6], proximity to waste recovery facilities, waste quantity and material specification [P_5] were highlighted as key factors to consider when creating end-markets for C&D waste.

While there are existing companies (i.e., Alex Fraser Group) that are already established in the market, there is a significant opportunity for more businesses to join them. This will then create the quantum and participants needed for a more sustainable market. The participants also pointed out that creating C&D markets can lead to extending the material's life and also reducing the associated GHG. For example, one participant used a concrete material example, whereby:

> "Looking at ways of reusing concrete to just extend that life, because it's got such a big impact on your greenhouse gas footprint, you want to extend that life as long as you possibly can. So, breaking it down and using it for aggregate in different paths, into road base is very important" [P_5].

Another example was shared about using buy-back policies and creating a contractual mechanism to facilitate this material transfer:

> "The other thing I'd be interested in understanding is buyback. Is there an opportunity for someone like Liberty Steel to buy back their steel once it's reached a certain life point? Or is there a way of creating a contractual mechanism that we say we buy X number of tonnes of steel, and then we put in a maintenance contract with them that says" [P_5].

A supportive regulatory regime [P_{10}, P_6] was also cited as being helpful to increase the uptake of such recycled products in the market, with participants stating that it would require careful formulation of specifications allowing PwRC in civil works and infrastructure projects.

Participants shared several best practices and successful organisations/ online platforms that guide waste trading. Planet Ark [P_6], WRAP and GIVIT [P_5] were some of the shared examples, and these are explained in detail in the improvement areas section below. They also highlighted the need for accessible tools to help industrial practitioners easily calculate the carbon footprint of waste material and identity CE opportunities.

6.3.2.3. Improvement Areas

The industrial practitioners shared a range of insights related to improvement opportunities regarding the creation and simulation of efficient end-markets in Qld. These areas include: (1) learning from global precedents and best practices; (2) aligning C&D waste management practices with the Qld waste strategy; (3) creating a CE through identifying waste material; and (4) economic opportunities through appropriate government incentives or disincentives.

Table 60. An overview of perceptions of market development for C&D waste PwRC in NSW

P	Intervention	Operation	Potential markets
P_4	Some practical examples were given of organisations that process waste in Brisbane, which show that the supply chain has responded to government drivers, particularly to the waste levy.	The waste levy has driven the price for landfilling significantly such that it now enables cost-effective recycling and reprocessing. If it were not for that waste levy, the market would go back to business as usual and opt for the waste landfilling option.	Steel, timber, metals, soil and concrete are the dominant waste types in a major transport infrastructure project. More opportunities can be found for plastics and cardboard depending on the volume of waste that is generated in the industry.
P_5	The markets are very material specific. The smart radius of the recovery sites and discoverability through the internet is impacting the growth of this market.	The current market operations revolve around specific materials. If it is not within a distance of easy transportation, that is very costly for the supplier. This can lead to very ‘hit and miss’ situations.	There are established markets for steel and concrete, especially considering that concrete needs to have a long life because it has such a large carbon footprint. While there are very specific markets, there should be more mechanisms to encourage buy-back policies through contractual mechanisms so that the residual materials are managed systematically.
P_6	While developing these markets is an effective way to manage C&D waste, challenges depend on the geographic location of the facilities. Some of the bigger regions have recycling stations and the ability to manage that, but more regional or remote facilities face numerous challenges.	There is some level of maturity on the basic main key streams of concrete and steel. There are more opportunities to grow markets in areas such as aluminium.	International best practices on low carbon emissions aluminium has 75% post-consumer recycled content within the product mix. This shows there are opportunities for more aluminium recycling in Australia. The top materials with potential markets for recycling are concrete, steel and aluminium, followed by glass. Furthermore, some of the semi-hard plastics could also be readily recycled. Take-back policies or takebacks for individual product streams are critical to reducing waste.

P	Intervention	Operation	Potential markets
P_7	Previous experiences of the local council's concerted efforts to recycle C&D waste and direct this material into road construction. With this evidence in mind, the creation of a C&D market is a strategic pathway for Qld to manage this waste stream.	While landfilling is still an easy and affordable option for some entities, connecting with new networks to negotiate where to send this material is necessary.	Crushed concrete has an established market and going forward plasterboard and treated timber can be expanded. It is also important to consider energy from waste options that are better than landfilling.
P_8	There are more opportunities to develop the market through updated TMR allowing more use of PwRC (i.e., glass).	As an end-market, there is a more accessible source of crushed glass to be available to asphalt manufacturers, which then catalyses those things becoming more economically viable and then 'business as usual'. The current economic situation is prohibitive as it cost more to put glass in than to not have the glass; looking forward, legislative and other guidelines will become more beneficial for such markets.	There are potential opportunities to fully or partially utilise materials such as crushed glass and plastics. There have been experiences in doing trials of plastics and car bumper bars as rubber in asphalt mixes. Different types of recycled bedding layers have been used in lower-level transport routes. There are future aspirations to try and use more PwRC in bedding layers rather than using the virgin clays and gravels in lower usage roads.
P_9	Intervention is a positive way forward aligned with Qld's waste strategy, specifically focussing on reducing the amount of waste going into landfills, creating a CE and building upon the subsequent economic opportunities.	While there has been an increase in opportunity and uptake concerning C&D recovery, Qld is still quite far behind other jurisdictions, such as Vic. The bottom line is the factor fundamentally driving businesses.	There is a great potential market opportunity for glass, timber and concrete. While there are mature concrete markets, the glass would lend itself very well to being part of that sector.
P_{10}	Developing a market for C&D waste products is a targeted approach for managing this waste stream and there are a few examples of early companies working in this sector (i.e., Alex Fraser).	In Qld, as the specifications around road-based projects allow PwRC, this has enabled growth in such markets. Moving forward as those specifications combined main roads adopting ISC requirements for major projects, there will be much greater demand for such products.	The quantities of recycling of concrete, asphalt and steel are significant. There are opportunities for materials like timber. However, the material requirements can be varied, through different types of timber, different ways and applications, meaning that matching of waste and needs might become a lot more complex.

Participants referred to several best practices and successful online platforms that guide waste trading. Planet Ark [P_6], WRAP and GIVIT [P_5] were some examples. P5 gave the example of GIVIT, where businesses can donate material that is not usable for them but valuable for others. Another participant said, *"Planet Ark and others for batteries and cartridges, there already are marketplaces that are really trying to support that, but it's not really joined up as an overall approach"* [P_6].

Two government representatives [P_8 and P_9] highlighted the importance of aligning C&D waste management practices with the three pillars of the Qld waste strategy. The first pillar is reducing the amount of waste sent to landfills through waste levies and mandated targets or mandated recycled content across a portfolio. The second is creating a CE through identifying waste material with potential for the secondary market and introducing market interventions to facilitate growth in those areas. The third is to build upon the subsequent economic opportunities through appropriate government incentives, or disincentives and cost competitive compliant PwRC. Participants suggested that, in alignment with these pillars, case studies and marketing around successful outcomes and where there is economic value could trigger the industry to opt for more renewable and circular processes.

6.3.2.4. Enablers of and Barriers to Market Development in Qld

This section presents key enablers and barriers described by the interview participants. As summarised in Table 61, there are myriad enablers and barriers related to the market development in Qld. The main enablers include improved regulatory frameworks with measurable targets for the industry; coalition between industry, government and academia; SP practices; incentives; understanding of material movement; and industry awareness on the value of secondary material. Furthermore, a funding model to demonstrate the value of using recycled material is essential to encourage more businesses to engage in C&D waste recycling.

One participant stated the incentives are critical to encouraging businesses to engage in the C&D waste market:

> "There might be some inconveniences, so you need the incentive to be there, which to my mind, the ones that are strong enough to offer that incentive are ISC and Green Star, and government and major private client mandates around the need to use recycled content materials and a significant percentage of recycled content materials" [P_{10}].

Table 61. Summary of enablers and barriers specified by the participants in Qld

P	Enablers	Barriers
P_4	• Product specifications enabling recycle quarry products in a variety of applications • The specifications need to enable a cost comparable fit for the purpose product • The supply chain has to deliver a safe and reliable supply of that recycled product • Procurement teams need to know to ask the right questions from suppliers	• Lack of precedents for new entrants into the market • Criteria requirements for virgin material over recycled material • Traditional procurement practices
P_5	• Quantum and location • Understanding of material movement • Contractual requirement allowing the use of recycled contents waste levies	• Chemicals and materials contamination • Timing and schedule • The locality of that second-hand C&D waste specific performance not meeting the standards
P_6	• High waste levies • More incentivisation that supports the use of recycled through a tax break or another benefit that actually provides an increase in the business value proposition • The push-pull market where material can be delivered and an increase in the requirements for recycled contents across all specifications • Valuing waste material as an input; that is, making a significant contribution to reducing both emissions and resource usage • A funding model that recognises the benefit without seeing it as just a cost impost • Alignment of improved regulatory frameworks • A more national approach for waste regulation	• The risk of asbestos contamination • The regulatory regime in place is unable to give businesses confidence around managing out the risk of asbestos fines • Inconsistencies of sampling rates across states • Limited post-consumer recycling is being supported at all levels of private sector and government to support • Insufficient facilities to segregate waste in sites • Waste contamination

Table 61. (Continued)

P	Enablers	Barriers
P_7	• Sustainable procurement • Setting directives, as in measurable targets, for industry and the reporting of those targets to make people accountable for their actions • Go beyond creating guidelines and focus on actioning • Case studies and marketing around successful outcomes and where there is economic value	• Perception that it is cheaper and easier to send waste to landfill
P_8	• Creating a coalition where the government is working with the private sector and the research sector • Amended policies • Building case becomes part of the accepted way of, in our case, infrastructure and maintenance activities anyway	• Strict procurement guidelines that more than likely would have been prohibitive • Difficulties in maintaining those engineering standards • Scepticism on whether PwRC are going to deliver the same outcomes or outputs as the virgin materials
P_9	• Economic incentives • Mandated targets or mandated recycled content across a portfolio • Pushing government departments to demonstrate long term, a percentage recycled content in all of their procurement and their projects • Appropriate project evaluation, to be able to consider a broader lifecycle perspective of the materials • Appropriate planning on how to use PwRC	• Complicated nature of a number of levers that need to be pulled to create the market (levies, planning requirements, stakeholder engagement) • Criteria requirements for virgin material over recycled material • Traditional procurement practices
P_{10}	• Reaching a quantum of waste and participants • Clear material specifications • Market incentive • Supportive government specifications, the industry frameworks, like ISC and Green Star	• Criteria that would exclude PwRC • Timing of material movement • Product testing and impacts on timing and logistics

The findings reveal that if C&D waste market practices are to be mainstreamed, there is a need for more relaxed specifications for using PwRC in construction projects. This will reduce the burden on the waste recycling industry for an actual 'fit for purpose' recycled material. Testing the use of mandatory minimum recycled content targets for government-led construction projects is highlighted as a targeted way for policymakers to lead by example. This will also encourage increased uptake of PwRC in any infrastructure project.

In terms of barriers, most responses were related to complex recycled product specifications, timing and logistics, plus risks of asbestos contamination and tight regulations. To overcome these barriers and to allow the growth of C&D waste markets, it is essential to reach a quantum of waste and participants.

6.3.2.4.1. Influential Factors for Materials End-of-Life Management in Qld

Several influential factors for materials end-of-life management were discussed by the participants. These include managing the perceptions about sending waste to landfills versus recycling [P_7, P_8], and incentives and data for evidence-based decision-making [P_9]. For example, one participant shared that localised recycling can be challenging:

> "Some of the bigger regions obviously have the recycling stations and ability to manage that, once you start to get more regional, or remote, then the ability to recycle becomes challenging and then you're in this world of landfill versus transporting it to another region to recycle it, which becomes difficult" [P_6].

Another participant added to this by highlighting the role of perception, stating, "*I think that there's a perception that it's cheaper to send waste to landfill, cheaper and easier to send it to landfill. I don't think I have to say anything more than that*" [P_7].

To better influence decisions related to end-of-life management of waste material, it is crucial to have more data. One participant suggested:

> "…it would be really interesting to use some of the data from Green Building Council and ISC, and look at the projects pursuing those ratings, which I think would be a logical application of these products. They are doing modelling of materials used up front, and they are also reporting

> on their waste, so you can get an insight quickly into what waste products are being generated and what demand there is for new products" [P_{10}].

6.3.2.5. Waste Regulations in Qld

The responses to the question related to current regulations and policies revealed a variety of aspects (Figure 38). Most participants shared their insights on relevant regulations with some comparison to other states. Overall, there were concerns around strict regulations, restrictive specifications and the critical need for approaches beyond compliance drive and demonstration projects led by the government.

P_{10} described the restrictive nature of the specifications in Qld, explaining, "*Particularly in Queensland it's tough because the specifications are quite restrictive, have held things back somewhat in terms of the ability to use a lot of recycled content materials.*" Another participant from the construction sector pointed out that the regulations should be able to give more confidence to the businesses, for example:

> "At the moment the regulatory regime in place is unable to give us confidence around managing out the risk of asbestos fines. Whilst we can use a certain thing, we can use recycled content for fly ash and glass and other things. Directly related demolition waste is a challenge because the sampling rates and then extend to which that's set up across the country, and each state is different" [P_6].

The current legislation tends to be compliance driven. For example, as one participant states about upcoming legislation: *"It'll be compliance-driven; legislation always is. I don't think it's going to achieve new outcomes, better outcomes. It's not going to challenge people to do better, it's not the right incentive at the stick"* [P_5]. Participants felt that, in future, there should be reformed legislation urging people to go beyond the compliance approach and achieve more advanced outcomes. A government sector participant, however, contended that the regulations are influencing more positive behaviour, especially with targeted guidelines for road infrastructure projects. For example:

> "I think they're currently quite good. Again, those TMR guidelines helped us. The waste levy coming in whenever that was coming up to a couple of years ago, that speaks for itself in giving industry impetus to minimizing waste. It's provided more guidance on direction, which is always good for any aspect" [P_8].

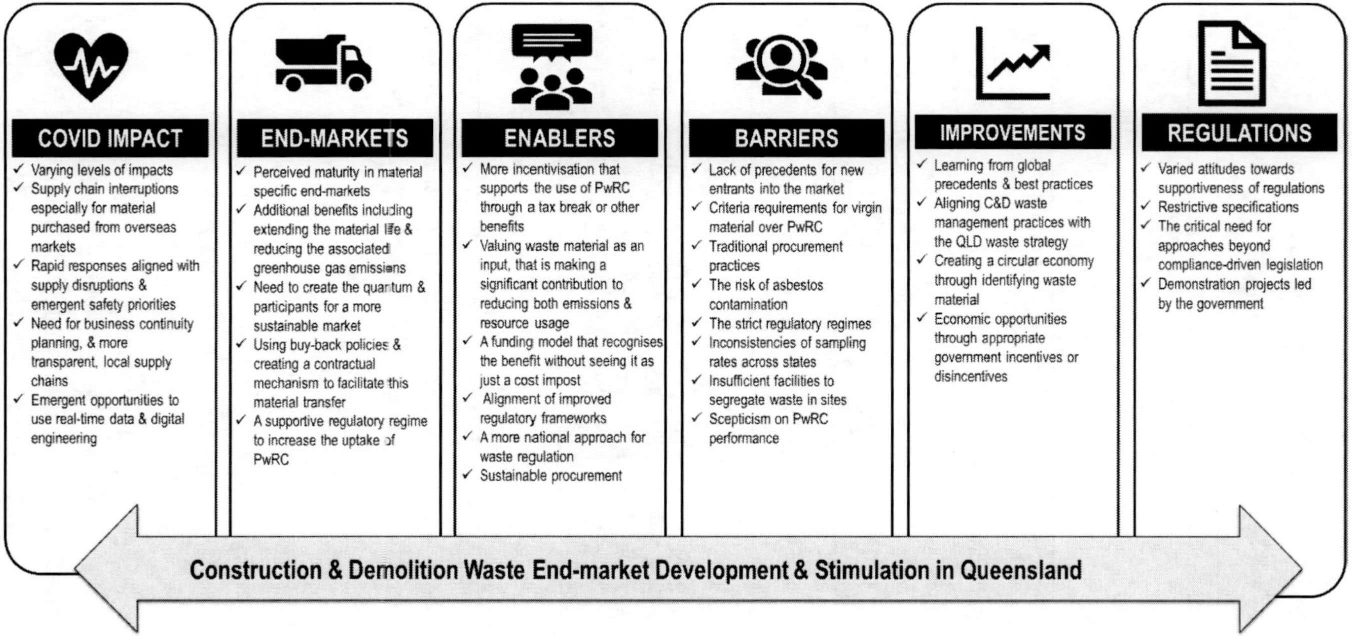

Figure 38. A conceptual model that describes C&D waste end-market characteristics in Qld.

Table 62. Summary of participants' opinions on the impact of current legislation on market development in Qld

P	Policy/Regulations
P_4	The government has taken actions to ensure the cost of disposal is more expensive than the recycling and reusing industry. Therefore, the value of those environmental and social externalities needs to be factored into a short-term economic aspect like the waste levy to support the industry. It is critical to show the wider economic benefits to make disposal at the landfill a last resort and thereby stimulating the market.
P_5	The challenge with the regulations is that it requires the bare minimum (do no harm) and limits the attempts to do better through better aspirations. It is compliance-driven, and it is not able to challenge people to do better, and therefore, it is not the right incentive.
P_6	The regulatory regime in place is unable to give confidence around managing out the risk of asbestos fines. While recycled content can be used for fly ash and glass and other material, material related directly to demolition waste is problematic. This is mainly due to the sampling rates and the extent to which that is set up across the country and differences in each jurisdiction. Incentivisation in some regions is much better than in others.
P_7	Previous local council experiences involved challenges in obtaining approval to use recycled C&D material in new infrastructure projects. Regulation has been a barrier for the industrial practitioners who are keen to move things forward.
P_8	The current regulations are generally positive. The Transport and Main Roads (TMR) guidelines and the waste levies gave the industry the impetus to minimise waste. It has provided more guidance on direction, which is always good for any aspect.
P_9	The current regulations address more because of a broader question around the waste. There is a lack of specific targeting C&D waste management, and there are more opportunities to create more comprehensive planning strategies and the development assessment and approval procedures.
P_{10}	There are challenges related to the restrictive nature of regulations (i.e., products specifications have limited the ability to use a lot of recycled content materials). There has been more progress in other states such as Vic, and to some extent, NSW.

Furthermore, there is a critical need for local government authorities to demonstrate best practices of using C&D waste PwRC in their large infrastructure projects and encourage increased uptake of PwRC in any infrastructure project. Table 62 summarises the participants' responses to the question on waste regulation in Qld.

6.3.3. SA

Only one interview was conducted in this state. The participant is the CEO of two recycling facilities based in Adelaide, which specialise in C&D waste resources and tyre waste recovery. At the time of the interview, the participant had been involved in recycling activities for about 10 years. The participant's primary responsibility was to manage crushing businesses and sell PwRC in various markets.

6.3.3.1. The Impact of COVID-19 and Responses in SA

The participant [P_{11}] stated that multiple impacts have been realised in the COVID-19 affected period (Figure 39). These include delays in delivering infrastructure projects across the state, delays between the C&D cycles and the inability to recruit a workforce. The first two impacts challenged the organisation to manage the mass balance of C&D required for the sustainable operation of the recycling facility. In the participant's experience, JobKeeper/JobSeeker schemes had the negative impact of labourers opting to take the government payments rather than doing hard physical work. This phenomenon has resulted in a significant increase in workload and the labour hour/cost percentage.

At the industry level, the participant stated that the industry and government worked well together in response to the issues that came with the pandemic. P_{11} had participated in several industry leadership calls with various state public organisations to communicate issues the industry faced during the affected period. The industry itself was also quick in taking preventative containment measures. As the participant described:

> "…that's what I think was interesting, I mean you look at it and you thought, and this was in normal circumstances, the entire industry would say they need three years to implement these types of changes and they happened in three weeks. So, for me that was a really interesting phenomenon that we saw right across the industry" [P_{11}].

At the organisation level, the participant indicated that their approach was to keep the workforce. However, the organisation implemented a 30% pay cut across their businesses to achieve this goal rather than laying off employees. P_{11} thought this approach was fair, understandable to the workforce, motivated them and enabled them to respond quickly to the impacts:

> "I think everyone felt quite together, everyone felt quite bonded by it because everyone was doing it together. And the people that performed more, the people that were in the higher income of the stream gave up more…as an executive team, we didn't have to start picking winners and losers. And I think that was incredibly motivating for our team, and I think it was incredibly motivating from our managers that, here is a company saying we want all of us to be in this together" [P_{11}].

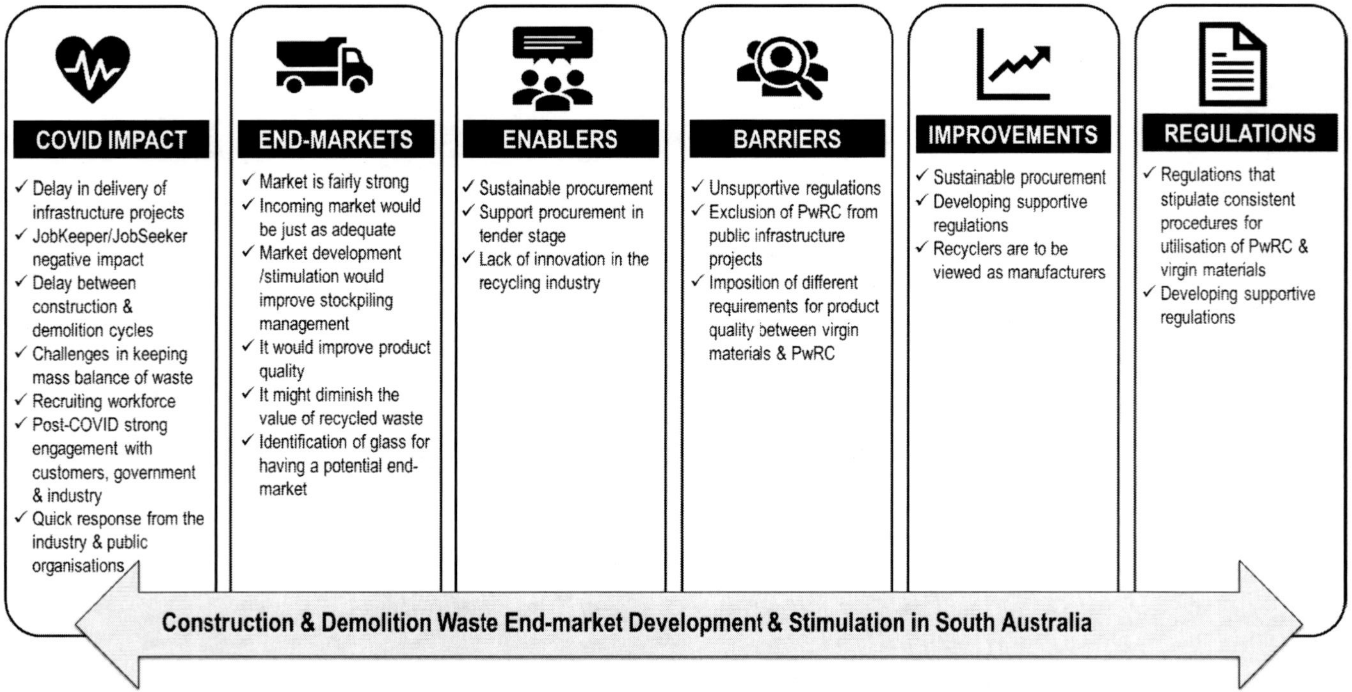

Figure 39. A conceptual model that describes C&D waste end-market characteristics in SA.

According to the participant, after the outbreak, the organisation engaged strongly with the government, the industry and their customers. This resulted in the organisation's ability to keep up with the market, which the participant described as "incredibly brilliant."

6.3.3.2. Improvement Areas

The participant expressed confidence in the resilience of the end-markets in SA and predicted a positive outlook for future markets as well. That prediction, however, came with two conditions: (1) enforcing strong regulation; and (2) the provision of an even playing field. On this theme, P11 indicated that developing supportive regulations is the key to create new and stimulate new end-markets. These regulations should be consistent across recycled and quarried materials and encourage sustainable government procurement of PwRC in civil construction projects. P_{11} states that recyclers are to be viewed as manufacturers, and through this lens, the value and quality of PwRC could increase.

6.3.3.3. Enablers of and Barriers to Market Development in SA

The participant's response on enablers for market development centred on efforts to activate the sustainable government procurement in buying PwRC and sending them to contractors. P_{11} suggested that tender documents should specify the requirements of buying PwRC for public construction projects:

> "…you could have governments buying, you'd have a whole range of quality control mechanisms that are in place to get onto the platform, and then you could have government something buying and directing to a contractor, such as, a big road project, that they have to go to the marketplace and acquire the recycled product and they could say, Okay, it has to be from this C&D marketplace, so therefore you know that you're driving sustainable outcomes from government procurement" [P_{11}].

Echoing participants in other states, to the participant, the primary barrier was also found to be regulations, specifically regulations that do not encourage the implementation of SP policy and exclude PwRC from civil construction projects. In P_{11}'s belief, government organisations such as the EPA view PwRC differently and impose more rigorous requirements on them than they do on virgin materials. Furthermore, the participant pointed out that innovation is limited as the general perception is that PwRC is not a manufactured product.

6.3.3.4. Market Development in SA

The participant indicated that the development of the market would result in better waste management, particularly in stockpiling and product quality areas, and have a perverse outcome through devaluation of commodities. In P_{11}'s opinion, market development will minimise the current variation in the quality of PwRC across the waste recycling industry in SA. As indicated by P_{11}, the negative outcome referred to a situation in which recyclers have not been deemed the manufacturer of a product, a factor that might discourage innovation from increasing the product's value. The participant stated that glass has a good potential for market development in SA. However, according to P_{11}, government policies challenge recyclers in approving their usage of recycled glass in roadworks.

6.3.3.5. Influential Factors Affecting Materials End-of-Life Management in SA

Regarding factors affecting end-of-life management, P_{11} believed that source separation is central to producing high-quality, less contaminated but cheaper products. In P_{11}'s experience, current on-site sorting has improved significantly:

> "Source separation now is much better, so the material we're getting is less contaminated. So, there should hopefully be less sorting and less cost around with that. And so, I think that should mean that we have greater ability, in terms of going forward, we should have more and more improved ability to produce products cost effectively and that are high quality" [P_{11}].

6.3.4. Vic

The participants based in Vic were representatives of two recycling facilities, one public (state government) organisation and one construction company (Table 54). Three of these research participants held management positions at the time of the interview, with one working as an environmental project advisor. The participants' exposure to C&D waste issues ranged between four [P_{13}] and 25 years [P_{15}]. The main focus of participants included R&D of

PwRC, market development and resource recovery in Vic [P_{13}]; waste management, crushing and landfilling operations, and technical management of C&D waste resources in NSW and Vic [P_{14}]; managing recycling of C&D waste, particularly asphalt in Melbourne and Brisbane [P_{15}]; and monitoring C&D waste disposal and use of recycled contents, and improving recycling waste generated at construction sites in Vic, SA and some contribution to other areas [P_{16}].

6.3.4.1. The Impact of COVID-19 and Response in Vic

In terms of the impact of the COVID-19 pandemic on C&D waste recycling activities, the respondents provided varying information. Vic had the greatest number and longest lockdowns in Australia since the beginning of the outbreak. These lockdowns slowed down the operation of some industry sectors. The impact and responses to COVID-19 are tabulated in Table 63.

6.3.4.2. Market Development in Vic

All participants agreed the development and stimulation of a market would be an effective measure to manage C&D waste sustainably (Table 64). Except for one participant [P_{15}], the participants could not evaluate the operation of existing markets for C&D waste sources. However, all four participants highlighted the necessity of improvements in the market. P_{15} mentioned that Vic's markets performance is good, and many other regions (i.e., other Australian states, the US, the EU and the Middle East) have been motivated by what Vic has achieved in using PwRC in civil construction projects. P_{15} linked these achievements to the better scale and economy of existing facilities in Vic that can produce high-quality PwRC. P_{14} explained that there are bad and good operators; the big recycling facilities with the largest market share receive negative feedback from authorities.

Regarding the potential markets, responses from two participants [P_{13}, P_{14}] underlined the potential market for glass and plastic in road-based projects. All participants stated that the markets for brick, concrete and steel are well established in Vic. Notably, one participant [P_{15}] mentioned that all given materials are already being sold and used in road-based projects. However, this participant highlighted the need to update the PwRC to add value.

Table 63. The impact of COVID-19 and the responses to the impact in Vic

P	Impact(s)	Responses
P_{13}	Work was not significantly impacted, but the organisation-funded research projects on C&D waste management were greatly impacted because of extended delays due to lockdowns, issues about budget allocation and uncooperative industry partners with changed priorities. However, the businesses working in the waste management and resource recovery industry were not impacted.	Not able to comment on this issue, due to not being involved in monitoring different procedures in response to COVID-19 within different originations.
P_{14}	The organisation was not impacted; to the contrary, it was quite busy serving many tier-one companies operating in Vic, such as level crossing activities at West Gate Tunnel and Metro Tunnel during the COVDI-19 era.	In terms of the industry response, on the individual level, this included measures such as social distancing, personal hygiene, taking care of one another and reporting if there is ever an issue with people have symptoms. Due to C&D waste recycling employees' unique socio-economic background, the need to repeat the message of the required practices and measures was essential.
P_{15}	The organisation was able to keep the operation alive, but some impacts were realised that were related to the market activity, which slowed down due to the economic impact of COVID-19. In the AEC industry, there was a lag in terms of the number of projects starting that is considered an economic slump.	The response was rapid and reasonably effective, as it was for lots of industries. The organisation measures included implementing many of the typical controls around shared spaces and customer interaction.
P_{16}	The level of impact on the industry was deemed to be great overall. The impact varied in different AEC sectors; while public construction projects continued during the outbreak, the situations with residential and commercial were far worse. Despite the government economic help (e.g., JobKeeper), the organisation had to reduce employees on some projects. Furthermore, the tenders for construction projects were also reduced. The organisation's ability to keep operating on public projects helped them to survive the situation.	The organisation's actions were very comprehensive due to the fear of being closed down. These actions, which had some cost implications for the organisation, included temperature testing in building sites, allocation of additional staff to each project called COVID monitors—whose job was to walk around the building sites making sure that the lunchrooms always maintained the minimum spacing and that social distancing was maintained, and the cleaning regime was adequate. The latter involved getting a sub-contractor to clean shared areas two–three times a day with the correct products per Vic's *Department of Health and Human Services* (DHHS).

Table 64. An overview of perceptions of market development for C&D waste PwRC in Vic

P	Intervention	Operation	Potential markets
P_{13}	A directory of waste materials has been developed in the organisation in the hope that they can better manage C&D waste.	While unsure about the operation of the existing markets, some stakeholders have given feedback that highlighted the need for support through infrastructure development, financial assistance or help addressing issues around compliance requirements.	Glass has a potential market as the organisation has received a fair few support applications for studying utilisation of recycled glass. The industry is investing in glass recycling facilities and the market, and there would be a huge market (demand) for the product, particularly in road construction. Furthermore, the opportunity for using plastic in asphalt is currently being explored.
P_{14}	The market that is facilitated by authorities would be helpful for C&D waste management, and it gives assurance that provides people with the confidence to operate within a market. To make this happen, a government incentive is needed to push market development for taking up PwRC in larger infrastructure projects.	No direct comment on this theme, but there is still a lot that needs to be done; for instance, Vic lacks a market and demand for recycled concrete. The current waste regulation regime does not support large producers of PwRC as opposed to other small operators.	The use of glass and plastic in low-risk concrete applications is a new market development opportunity in this space. As for brick and concrete, their aggregates can be fed back into the concrete production.
P_{15}	The market proves to have positive performance in Vic and has been an excellent intervention to manage C&D waste materials better. However, in other states, it is underperforming. The markets in other states are driven more by road agencies than environmental agencies. Through sustainable procurement, road agencies can encourage using PwRC in civil projects.	Vic is leading PwRC in civil projects in Australia and overseas. The advanced facilities have been able to produce high-quality materials. However, its current performance is being jeopardised by the introduction of new specifications limiting the use of PwRC.	There is a huge market for all materials listed (brick, concrete, glass, steel and asphalt), and in Vic, they have been sold already for road-based projects. However, an update on the product is needed to add value to these materials. Concrete, brick and glass have more potential jurisdictions like Qld.
P_{16}	The market development will bring huge benefits to managing C&D waste materials as long as poor (contaminated) waste disposal is avoided.	Understanding of existing markets is limited as there is no benchmarking system to measure the operation of the market.	Limited knowledge of the market for the given materials. Concrete and brick are routinely being used in road projects, and steel is continually being recycled in Vic.

6.3.4.3. Improvement Areas

Regarding improvement areas for the creation and simulation of efficient end-markets in Vic, participants had different opinions; this is primarily linked to the needs and requirements of the sector (government, recycling or construction) they represent. However, the input from the participants collectively can help shape the bigger picture. In the government sector, one participant [P_{13}] outlined three improvement areas: (1) development of knowledge of feasibility of using PwRC in different applications; (2) investment in recycling facilities through government support (i.e., Recycling Victoria Fund); and (3) decentralisation of recycling facilities to service the entire state.

Participants from the recycling sector [P_{14}, P_{15}] provided a well-balanced response on improvement areas that correspond to education, encouragement and enforcement. P_{14} recommended attitudinal changes of major stakeholders (i.e., councils, asset owners, *VicRoads*, *SV*, etc.) through raising their awareness about PwRC. The participant suggested that opportunities such as roundtable discussions could assist in achieving this aim by communicating information focused on sustainable outcomes (e.g., carbon footprint reduction), financial incentives and the competency of PwRC compared to virgin materials. P_{14} anticipated that such information sharing would facilitate decision-making towards allowing and sharing the risk of using PwRC among large asset owners (i.e., VicRoads), a phenomenon that would be followed by contractors working in parts of large projects and therefore enabling significant PwRC uptake in Vic.

P_{15} highlighted the need to build new and preserve well-operated recycling facilities close to metropolitan areas (markets) and to apply stringent enforcement on all recyclers (i.e., registered and unregistered). While the former factor provides a financially competitive advantage over the virgin materials shipped from long-distance quarries, the latter would counter the negative image of PwRC created by poor processing performed by dodgy waste recyclers. This participant also pointed out that their organisation had waived the gate fee, which had made recycling a viable option for waste producers and encouraged site separation. They explained that, under such circumstances, a large-scale feedstock is made available that can compete with virgin counterparts.

P16, working in the construction industry, emphasised regulation, accreditation schemes and uniform reporting systems as useful measures. The two first strategies would motivate a construction company to prioritise waste management during the planning stage. Notably, for accreditation schemes that reward sustainable waste management practices in the industry, the participant believes that customised but national rating and reporting systems can significantly improve management best practices.

6.3.4.4. Enablers of and Barriers to Market Development in Vic

This section focuses on the major enablers and barriers specified by the participants. As summarised in Table 65, various enablers and barriers are reported. The main enablers in Vic were found to have a diverse nature, including encouragement factors (the development of infrastructure, supporting local government to buy PwRC and a central waste marketplace), education (confidence building, R&D, and demonstration projects) and enforcement (supportive and simple policies and specifications). In terms of barriers, all responses indicated that limited knowledge about PwRC could hinder their application in construction projects. Furthermore, except for P_{13}, who referred to higher costs of PwRC, none of the responses involved technical or economic barriers. Interestingly, the two recycling representatives argued that PwRC produced in Vic have comparable quality and are offered at lower prices compared to virgin materials.

6.3.4.5. Influential Factors for Materials End-of-Life Management in Vic

The participants' views on influential factors provide insight into the mechanisms underpinning decisions about and behaviour around end-of-life management in Vic (Figure 40). The factors specified by participants included a mix of economic and technical issues, including the scarcity of virgin materials, the increasing cost of extraction and sustainability programs of various organisations [P_{13}], potential and limitation to recycling a waste resource multiple times [P_{14}], the closeness of production to market and the extent of the economy of scale [P_{15}], and limited incentives and investment in innovation in the recycling sector [P_{16}].

6.3.4.6. Waste Regulations in Vic

The responses to the question related to current regulations and policies revealed different aspects. While one participant [P_{14}] could not provide a comment on current legislation, the other participants had various opinions ranging from positive to negative perceptions of regulations and specifications [P_{15}], to making overall improvements [P_{16}] and ensuring the guidelines and policies materialise in action [P_{13}]. Table 66 summarises the participants' responses to the question on waste regulation in Vic.

Table 65. Summary of enablers and barriers specified by the participants in Vic

P	Enablers	Barriers
P_{13}	• Development of recycling infrastructure • Support research and development • Build confidence for recycled products in the market • Support local governments to use recycled content in infrastructure projects • Development of policies such as Recycled First Policy	• Higher cost of recycled products over virgin materials • Limited knowledge about possibilities for using PwRC
P_{14}	• Assurance for using recycled products • Create government-backed successful demonstration projects for learning and promoting purposes	• Negative perception towards using recycled products • Lack of education and demonstration of using recycled products • End-users being risk averse
P_{15}	• Develop straightforward specifications for recycled products	• Lack of comparative studies investigating recycled products quality against virgin materials • Double standards when comparing virgin materials and recycled products • Unreasonable ban on using PwRC
P_{16}	• Develop a central marketplace that advertises current and upcoming C&D waste • Develop a credit program to encourage ethical waste management, including uptake of recycled products	• Limited knowledge about available waste and recycled products

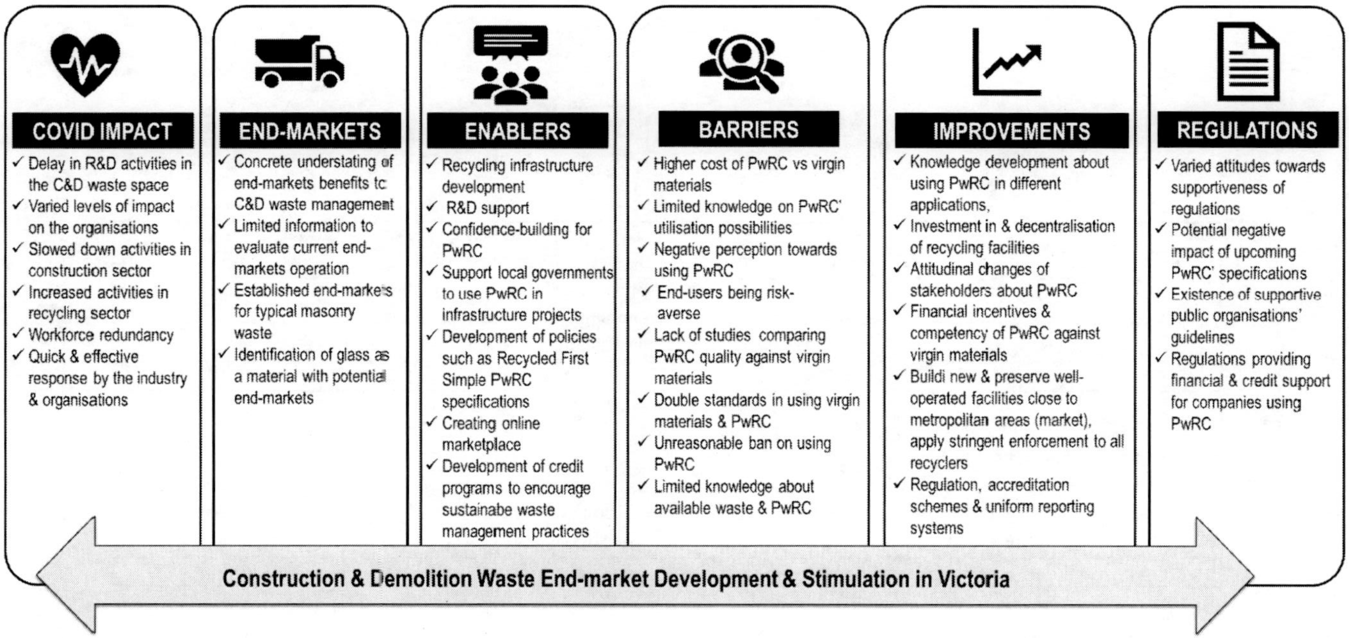

Figure 40. A conceptual model that describes C&D waste end-market characteristics in Vic.

Table 66. Summary of participants' opinions on the impact of current legislation on market development in Vic

P	Policy/Regulations
P_{13}	It is not clear how the new EPA act released in July 2021 will be impact the market development. Referring to one recycling company in Vic, the current regulations are unsupportive. Sustainability Victoria has developed policies that assist stakeholders in further developing markets through incentives such as financial support for infrastructure development and/or sustainable procurement. The Victoria Recycling Policy encourages the increased use of PwRC to drive the market. Other organisations such as Major Projects Victoria have also developed guidelines. A clear roadmap showing how to put these guidelines into action is still lacking.
P_{14}	No answer provided.
P_{15}	The current legislation is adequate but new specifications will negatively impact the use of PwRC across Vic. Mandating minimum recycled content in regulations is not supported but instead stakeholders should be informed and helped to identify opportunities to use recycled content.
P_{16}	Improvements are needed, and one example of this is to accredit companies using recycled content or providing monitory incentives through government leadership.

6.3.5. WA

In WA, 11 participants representing various stakeholder groups in C&D waste space were interviewed. Five of these participants work in government organisations, three represent the recycling industry, one provides waste-related consultation services, one operates in the construction industry and one was a technical expert in the construction materials manufacturing industry. As outlined in Table 54, nine participants held management positions at the time of the interview, one was a principal engineer, and one was a principal advisor. The participants' exposure to C&D waste issues ranged from four [P_{25}] to 19 years [P_{26}]. The main activity of participants includes consultation for and management of C&D waste in the construction industry [P_{17}, P_{22}, P_{25}]; recycling C&D waste resources [P_{18}, P_{21}]; government-led monitoring deconstruction projects for residential development sites [P_{19}, P_{20}]; manufacturing construction materials [P_{23}]; and contribution to delivering the state waste strategy objectives and market development [P_{24}, P_{26}, P_{27}].

6.3.5.1. The Impact of COVID-19 and Responses in WA

Not all respondents could describe what the market development would look like after the COVID-19 pandemic. Participants from the recycling industry did not see any difference in market development before and after COVID-19 [P_{18}], which was also confirmed by two participants representing government organisations [P_{19}, P_{20}], who argued that the lockdown was minimal. Construction is booming in WA, and the market for recycled material is seen to be viable before and irrespective of COVID-19. However, two government representatives [P_{24}, P_{27}] mentioned that due to the government recovery plan in response to the COVID situation, including bringing infrastructure projects forward to support local industry, the market would be contributing more to C&D waste streams after COVID-19 (Table 67).

6.3.5.2. Market Development in WA

Market development in WA is a challenging topic as it is seen differently by stakeholders representing different entities. In Table 68, the opinions of 11 WA participants on market development as an effective intervention for better C&D waste management, operation of the existing market and the potential market for certain C&D waste materials are explored.

Table 67. The impact of COVID-19 and the responses to the impact in WA

P	Impact(s)	Responses
P_{17}	COVID is not really an issue that has affected the uptake of recycled crushed concrete in WA. There was an initial shutdown, but it did not affect the industry.	No clear answer provided.
P_{18}	As the waste operation is considered an essential service, the operation had not been affected. Furthermore, self-isolation among the workforce in operation is a norm that minimises the impact. Lastly, incentives in the residential market and not commercial has been counteractive to this impact. In WA, it has been fairly unchanged. Because of the fact that so much civil work is done, businesses can conduct themselves with the right protocols for COVID.	Developing a few protocols in response to COVID-19 situations.
P_{19}	There was some disruption but overall minimal impact on activity, and we have managed to keep things going as the state experienced minimal lockdown.	All the stimulus in terms of housing and economic stimulus has actually increased activity. Generally, the pressure is probably being mainly on the construction industry. Typically, they have had excess materials as opposed to shortages. The extra construction work undertaken is actually used up to stock balls or to have some of those raw materials turned into recycled blocks or recycled road-based materials.
P_{20}	Shortage of resources, both human resources and shortage of materials, particularly steel and timber. Probably pressure on industries that had probably wound down a fair bit, as the state was in a fairly flat market where there had not been much growth, and actually, builders were probably looking for work, whereas now, they have to turn it away. There were some changes regarding how many people could be working on construction sites. For example, particularly building houses, there were limits on the number of trades inside the homes, etc. However, that probably, if anything, just exacerbated the general problem with lack of resources and slowing the construction down.	No clear answer provided.

P	Impact(s)	Responses
P_{21}	COVID has had no real impact on the company from a construction perspective.	During the outbreak, the company went to the memorandum of understanding to share sources if the situation worsened. There were many releases on planning caveats around the COVID-19 period to try and stimulate the economy, in other words, get people spending money on project planning, which meant a lot of programs suddenly came live in the planning baseline, which overwhelmed the local government approval processes.
P_{22}	COVID-19 has not affected the company's work routine.	Application of the segregation and the work-at-home requirements advised during the lockdowns. The company has staggered staff 50% home, 50% in and following all measures that were recommended.
P_{23}	The company has not been impacted on COVID-19 at all. At the moment, the company is busy such that it has just not had the time, people and resources to keep up with the boom in the construction industry.	The company follows the advice from the state health authorities, such as social distancing, using face masks. The company staggers the times for people to have breaks, X amount of people are allowed to be in a lift.
P_{24}	It had a significant economic impact, and there is probably much pressure on our organisation and on the industry to do better, to recycle more C&D. It is probably partly to do with COVID, but it was a priority before then anyway; it just so happens that some of these infrastructure projects that provide an opportunity to use more material have been brought forward. So, there is even more pressure, probably than there previously was.	The government's response is to try and fast-track infrastructure projects as part of the economic recovery. Those major infrastructure projects provide an excellent opportunity for the government to increase the use of C&D waste PwRC.
P_{25}	From a C&D point of view, obviously, the onsite work stops when there is an outbreak, which negatively impacts the project programming and delivery. However, the residential housing market is doing well, so there has been a positive impact in the broader macroeconomic perspective, and the company is busier than before. At the construction site level, COVID-19, probably more in 2020, definitely caused significant delays that created significant uncertainty within the business, but it seems that it has rebounded in 2021 at the moment in a very positive way.	Overall, the industry made a good effort to respond to COVID-19. Working from home and preparing for the booming residential market were strategies that the company undertook to respond to these situations.

Table 67. (Continued)

P	Impact(s)	Responses
P_{26}	COVID-19 has not had a significant negative impact on the use of C&D products in the organisation as the construction industry was designated as an essential activity from the very beginning. COVID has actually led to an increase in the organisation's work. However, the recycling industry suffers from manpower and equipment shortages and lacks appropriate alternative plans, which makes it difficult for them to keep up with the amount of work.	Both the federal and the state governments are using infrastructure construction as a stimulus for the economy. Therefore, the organisation is doing record amounts of work, within the organisation; working from home where possible was the most common answer to this situation.
P_{27}	The impact was more on the change in the priority of the organisation's activities, such as bringing forward infrastructure projects to make sure the economy keeps on ticking over, and there are local jobs and local industry development. Furthermore, skill shortages and capability shortages have made it difficult to procure materials and services quantity and cost-wise in terms of procurement.	The industry has responded positively to keep business going and implement processes and practices to ensure that operations continue in a safe environment and still be productive. It has brought in hygiene habits and so forth, and contact tracing. The government responded to the COVID-19 situation by supporting the use of recycled content and increasing the number of projects, which has broadened the opportunities.

Table 68. An overview of perceptions of market development for C&D waste PwRC in WA

P	Intervention	Operation	Potential markets
P_{17}	Creating markets is absolutely essential, and currently, recyclers are stockpiling PwRC as there is a limited market for their products.	The people in the industry are doing their absolute best to try and market this material. However, there is some reluctance among councils that are the main end-users to procure PwRC that have exact quality as virgin materials and meet the specifications. Furthermore, MainRoads WA currently promotes using PwRC in low-value applications, which should be changed.	There is a potential for new markets and existing markets in brick, concrete, glass and asphalt. Recycled concrete and brick in the state have similar performance as new concrete and have a good market. There is a little bit of an issue about glass with the alkalized selector reaction. There are issues with brick pave roads' performance, and they have to be converted to asphalt. The market for steel can be improved by legislation that stops the demolition of steel structures and rather keeps it intact for reuse.
P_{18}	Without a doubt, the development of the marketplace can result in better management of C&D waste in the state. However, the focus of government should be equally on waste handling and PwRC. With government support, PwRC create a market for themselves.	The existing market operation is very poor, and it can improve by benchmarking it with other states (i.e., Vic and NSW) and overseas markets.	Steel is an exception because there are already many markets for that, and because that market has been created, there has not been a huge demand to divert. Steel is handled well, as far as recycling is concerned. But certainly, concrete, brick, glass and asphalt can have a potential market. Road base as an end product has the biggest potential here to have an impact, followed by clean fill-sand. There is an issue with recycled sand as its virgin types are abundant, and recycled sand is not as appealing to look at as virgin material. Asphalt is gaining some ground in reusing materials in asphalt production. Glass is the biggest tragedy. Glass, despite having so many fantastic properties, still is not acceptable by government organisations.
P_{19}	The market for C&D PwRC exists in the state. However, they are not as widely known as markets for fresh materials.	The operation of the markets is good and growing, but still, the drivers lack in terms of procurement because it is too easy for people to see access 'business as usual' materials and we have plentiful supplies of those as well.	Brick, concrete, glass and sand have existing markets, but in terms of the ones that probably do have the most potential for the future, in WA those are steel, glass and asphalt.

Table 68. (Continued)

P	Intervention	Operation	Potential markets
P_{20}	The elements for the successful development of the C&D waste PwRC market already exist in the state. However, fresh materials are being ordered routinely. Therefore, the development of a market is a good intervention for waste management if the demand is created for such products. Mandating the use of PwRC and educating engineers and people who can choose between PwRC and virgin materials is needed to stimulate the market.	Market operation is good but can be improved by providing education and information that motivate people to use recycled products at a higher level.	There is no recycling facility for glass in WA. Asphalt is getting reused, but certainly, recycled steel and recycled glass are probably two opportunities for market development. The other materials like bricks and concrete that can be crushed tend to be put to use for fairly low value purposes.
P_{21}	With regards to creating a marketplace, the marketplace is already here, and the action needed to solidify that marketplace is to remove risk from the outbound products. However, the ambiguity around the definition of waste in the state, the unwillingness of people to buy recycled products due to risks associated with the definition of waste versus resource, transportation cost, complicated testing requirements and inconsistent testing requirements for fresh and PwRC, plus location of the cost,	Some very good players have been in the space for a long time and have developed a good name. Some recyclers in the state have produced high-quality materials by implementing strict criteria, improving machinery and processing arrangements and getting assessed by independent auditing entities as a government mandatory requirement. The audits provide an open platform that will have a certificate of quality of recycled products (structural and environmental wise) that reassures buyers. The product that the participant's organisation produces gets sold with low margins, which justifies the company's processing costs.	The market for sand, concrete and drainage rock is established, and the participant's organisation is selling these materials. There is a limited market for glass, which can be provided depending on the clients need. However, it is cheaper to be imported from China than manufactured or repurposed in the state, given the high energy costs. Notably, glass from C&D waste streams is toughened, which causes issues in particle size distribution. Timber does not really have as much value; it is more of a problem stream in WA because of contamination of products like medium density fibreboard will eventually become the asbestos of the future, because it will flux and does not burn, so it has no value in an incinerator, either. The problem with brick is that it gets mixed with other materials such as timber, making the recycled product a very poor grade material.

P	Intervention	Operation	Potential markets
	can determine if market development can achieve the waste disposal reduction objectives.	However, some companies do not apply strict processing practices in the market, and their recycled products do not meet specifications used in the state. The current market needs to be set up so that suppliers can find clients and their expectations in terms of testing requirements and predict the demand for their product. This can be complemented by implementing a forward supply chain and lifecycle assessing the demolition and construction projects to anticipate the amount of processing waste. This kind of arrangement does not exist in the state for several reasons, including the change of governments and programs thereof and the lack of knowledge of where to build recycling centres. The latter is important to keep up with the pace of urban growth to ensure that end-markets are not affected. The operation of the existing end-markets in WA is affected by the lack of sustainable supply of waste materials to recycling facilities with good practices, as waste may go to other companies that make basic crushing and send it to landfills.	
P_{22}	Market development and stimulation is definitely an effective strategy to manage C&D waste sustainability if set up properly.	There is probably no known marketplace for recycled products. It is a bit focused on perhaps one or two companies who do it and that they can use the concrete or the bricks or whatever. There is no known market for other entities to go and obtain those materials.	Steel, brick, concrete and glass definitely have a potential market because they are the bulkier items and are not hazardous.

Table 68. (Continued)

P	Intervention	Operation	Potential markets
P_{23}	The market development is definitely an intervention to create and stimulate the market for C&D waste PwRC.	The operation of the market is good, but currently, it is all low-value products. However, the demand for recycled content in manufacturing construction materials should be high and consistent across end-users.	In WA, concrete is recycled routinely, and whatever waste concrete comes back, recyclers divide them into aggregate and sand, which can be reused. Apparently, the state does not have the manufacturing to convert glass waste into a usable product for glass.
P_{24}	It probably is an intervention and certainly is critical, and that applies to any recycled material. Recycling does not work unless you have a strong market for the materials that are being recycled. So those materials need to be produced to a particular specification, be cost-effective compared to virgin raw materials, and protect human health and the environment. The government, through different public organisations, is supporting the market to address these issues.	The operation of the market is affected by the inconsistency of waste operating companies. Some of them are proactive and very keen to access markets for government projects, and they are very keen to make sure that their material meets specifications, and others are probably less keen and probably less about making a quality product, and perhaps more about just finding a pathway for the material. Therefore, currently, there are many differences in the recyclers and the materials they were producing to reach different end-markets. In the state, government organisations working together to instil confidence in C&D waste PwRC by having various initiatives, chief among them is the Roads to Reuse Program. This program provides environmental standards for C&D waste PwRC, notably in road applications, supporting recyclers to conduct regular sampling and testing, and running an independent audit program.	Steel is pretty much extracted from the stream at the point of drop off and sent straight to recycling markets. Glass, unfortunately overly in WA, there is no recycling facilities that recycle bottled glass back into the bottled glass. Most of the glass that you would find in the C&D stream is different sorts of glass, and there are opportunities to use glass for pretty low value uses such as civil construction. If high-value uses and economically feasible higher-value uses can be found, that would be great, but at the moment, we are struggling to do that. In this organisation, concrete is the focus simply because it makes up a significant proportion of the C&D stream. As for brick and asphalt, there are opportunities for new markets, but that probably goes back to the scale.

P	Intervention	Operation	Potential markets
P_{25}	A fundamental driver for reducing waste is the cost and also just that it is the right thing to do. If there was a better way to actually utilise materials that still had significant value and avoid them needing to be put in a bin and trucked offsite, if there was a way to recapture some of that value, that is a pretty major initiative or incentive to reduce waste.	I have limited knowledge of the existing market, but my organisation attempts to look for opportunities to use PwRC if it is not expensive, does not have practical issues and is not difficult to use.	There is a very low market for glass given the state is transitioning to high-performance glazing systems, and the opportunity to recycle glass in terms of glazing has been quite low, given the glass scratching. For brick, there is a really high potential, and it stands out, as it is aesthetically nicer in many respects than new bricks; however, the costs for the supply and removing the lead paint and asbestos is a barrier. Recycled timber is quite difficult to include in a new building; even though it might look nice, there are all sorts of issues with timber warping, including legal issues. Structural steel is probably difficult, given its requirements, if it is damaged and if the structural qualities are maintained.
P_{26}	It is an effective intervention and organically and already exists. However, the issue is to develop the demand for the product or some applications. Creating electronic web-based marketplaces for C&D Waste in WA is needed for people to exchange their waste and PwRC.	Every asset owner has their own opinion, and obviously, that influences their policies and standards, and the result of that is it influences their appetite for the use of C&D waste. For instance, although having the same minister, the MainRoads WA is more progressive with the use of C&D waste than their portfolio partners, the *Department of Transport and Public Transport Authority*. Therefore, the market is really driven by perception and historical practice.	There is an established market for asphalt, and it is a 'business as usual' product in WA at the moment. However, the volume of the market uptake is not as large as in eastern states, i.e., Vic and NSW. The market for recycled concrete is growing, but additional improvements are needed to further uptake this material. There is no major recycling facility in WA for glass, and the only option is to send it to SA, which does not occur, so finding an application for glass is important. Some processors, however, convert the recycled glass into sand, which is not preferable due to the abundance of this material in the state. There is some limited waste trading for timber, which is upcycling timber to mulch. Recycled sand has a flourishing market and still has a historically established market due to its economic value.

Table 68. (Continued)

P	Intervention	Operation	Potential markets
P_{27}	It helps to manage C&D waste in the state. As soon as you place a value, and you stimulate a value on certain materials, it helps form an industry and form the economy needed to sustain it. However, an impediment to forming this market is economy of scale. The organisation is assisting in creating the scale by investing resources such as setting an ecosystem in which buyers receive confidence in using recycled products that are safe from an engineering and health and safety perspective.	The state is slowly starting to establish end-markets. MainRoads has had to champion the establishment of two key end-markets regarding crumb rubber and crushed recycled concrete, the establishment of environmental specifications and independent auditing when it comes to following the specifications from the industry. The participant's organisation required that that process be embedded first before they committed to the procurement. So, it took a little while to work cross-industry and the other regulating agencies to establish that. Probably the one thing that is still an issue is leakages from the landfill levy system at this point in time because there is not a consistent approach, as some C&D waste can be trucked outside the levy area. It creates a scenario where there is leakage from the catchment of the levy, and then also the materials are going to different landfill operators. Therefore, it is not coming back into the system, which needs to be overcome in further changes.	There is an existing market for concrete, asphalt and clay. In this organisation, clean fill and recycled sand have a special place for application in road projects. Glass has limitations for use in asphalt and concrete, and therefore it is better if the glass is recycled back to glass. The issue with potential markets for these materials is that the industry is not developed, and there are not many buyers, including a local government that does not support the state government in sustainable procurement efforts. Furthermore, the PwRC are mainly used in big projects, and smaller projects cannot use these materials.

6.3.5.3. Improvement Areas

In response to the question regarding improvement areas to create and stimulate end-markets, participants highlighted a wide range of areas that need immediate attention and planning. Most of these areas are related to the role of the state government and publication organisation in driving the improvements. P_{17}, for instance, mentioned the lack of collaboration between end-users (e.g., government organisations) and the research and development sector on leveraging the findings obtained from trials and field experiments by government organisations and industry in large-scale applications. To foster this, the participant suggested that behavioural change in the industry and government should be pursued:

> "Because there's nothing wrong with the products, nothing wrong at all, it's the people's reluctance to use them…I think that's absolutely essential and if universities can work out a way to change people's attitude towards this product, you know, that's the only thing that I can see that's gonna work now" [P_{17}].

P18 mentioned the government leadership role in stimulating the market by accepting PwRC and using minimum recycled content. P19 pointed out the government's contribution, such as improvement in procurement policy and specifications for PwRC. P20 indicated that improvement areas should be focused on enhancing authorities' understanding involved in approving developments of recycled product alternative materials to agree with their applications in residential developments. For example:

> "Better specifications for recycled products. That would go a long way in terms of overcoming some of the resistance in the industry, if we proper risk perspective. The people can be assured that what they're getting is of a suitable calling for the task" [P_{19}].
>
> "…standards and the specifications probably need to be brought into play to recognize the need for the use of recycled materials" [P_{20}].

P21, who represented the recycling industry, emphasised improvements in the definition and classification of waste and implementing an alternative approach to the current case by case analysis and approval of manufacturing waste products. This participant also suggested that providing support for recyclers who need to pay for rigid testing regimes for PwRC should be on the radar for further improvements, explaining:

> "I myself see that (Roads to Reuse Program) as being a positive step, but nobody has done economic modelling on it, and as it stands currently, it makes it financially unviable to sell that product. It increases the cost too much for us. Why sell a product that we've had to pay an extra $2–4 a ton on testing, maybe even double or triple the cost of the material, when we can sell it now and we've got to clear our stockpiles" [P_{21}].

P_{22} and P_{23} identify the need for improving the government leadership role in raising awareness of the benefits of PwRC. This awareness can take different forms such as recognition, uptake and customised recycled product specifications. According to the participants, education plays an important role in the latter form, as it gives knowledge to engineers who write specifications and breaks the habit of being risk averse:

> "I think more government uptake on what are the goals of what that process is trying to achieve. So, government recognition and government advice of the community benefits for that, and awareness in the industry of the availability of certain products through that process. There is no current awareness. So, a two-staged prong government, awareness, and facilitation, and industry knowledge that it exists" [P_{22}].
>
> "So, we're going to get the government on board, they themselves, they should be leaders in this field, but they're the opposite" [P_{23}].

As indicated by P24, providing clear guidelines, similar to the Roads to Reuse guidelines that offer a whole range of different products and end-uses, is a step that government needs to take to stimulate the market for C&D waste PwRC. These guidelines help the purchaser to make more informed procurement decisions:

> "I think that the purchasers need more and more clarity and support for their own procurement. So, they want to see what's the guidance, what can I use the materials for, and have confidence in both the engineering aspects but also the environmental aspects so that they're confident that they're not buying materials that have got asbestos or something like that in them" [P_{24}].

Furthermore, this participant highlighted the issue of transport cost and maintained that PwRC should be available close to where it is used. P_{25}, who commented on the improvement areas from the construction industry perspective, mentioned two major issues: a design for disassembly and a strong supply chain for PwRC. According to P_{25}, design for disassembly thinking helps with the quality of PwRC supply chain, making waste materials such as engineered timber much easier to be processed. Regarding supply chain, in P_{25}'s experience, issues with recycled product supply chain are profound, including the inability of suppliers to provide materials in the quantity and quality required for the construction industry within specified time restraints:

> "Our subcontractors aren't set up to supply the quantities and the volumes, and what have you, it's a supply chain issue, potentially. I don't know, but anecdotally when we want to do recycled materials it's more expensive, they're generally harder to get, it doesn't fit into our typical project timeline as easily, so there is project risk associated with it, whereas new materials, it all just fits together nicely. So, I don't know, that's my crack at answering that one" [P_{25}].

P_{26} and P_{27}, being government representatives, indicated the need for improving understanding of PwRC characteristics, such as providence; dealing with materials falling outside of levied system; and establishment of more recycling facilities supplying PwRC. These were highlighted as major improvement areas to facilitate creation and stimulation of efficient end-markets for C&D waste materials. Further, participants suggested that how the waste is generated, handled and processed needs to be known to those who will use it, including government organisations. This way, the confidence needed to make purchasing decisions becomes available because:

> "…as soon as you use that 'W' word, nobody wants to know. So, it's really about maybe providing better visibility of what actually C&D looks like, where it comes from, how it's handled, and the very high standards that you're able to achieve" [P_{26}].

Participants contended that dealing with material outside of the levied system (waste material leakage) would improve the state's resource efficiency and CE, given the quantity of these materials that could be captured back into the system and reused. More recycling facilities with suitable capacity can meet a potential demand created for PwRC across the state. Therefore:

"There needs to be the establishment of more suppliers of the materials. So, so far, we've got a limited number of suppliers of the material in a small area of the state, supplying the materials to achieve broader use, and broader uptake, we would need more suppliers to come online to either treat crumbed rubber and manufacture crumbed rubber to the specifications that we need" [P_{27}].

6.3.5.4. Enablers of and Barriers to Market Development in WA

Enablers of and barriers to market development and stimulation were identified in the interviews with WA participants. Table 69 summarises participants' responses regarding the major challenges and opportunities for market development in WA.

6.3.5.5. Influential Factors for Materials End-of-Life Management in WA

Regulation was frequently cited by the respondent during interviews as a factor influencing C&D waste end-of-life management in WA. For instance, P_{17} indicated that as the waste materials need a lot of testing and quality control, people are less interested. This participant also mentioned that specifications (i.e., Specification 501) developed by MainRoads WA considers waste materials as a base course material. This has resulted in local governments being reluctant to explore other possibilities of using PwRC such as those implemented in other states (Vic and NSW). P_{19} also confirmed that the state waste regulation is a hurdle to effective end-of-life management in WA. P_{27} indicated that the lack of legal definition of waste in regulations limits sharing certain waste resources between different construction projects.

Uncertainty about the waste lifecycle was also mentioned by multiple participants [P_{18}, P_{19}, P_{20}, P_{21}, P_{24}, P_{26} and P_{27}] as having a great degree of influence on waste end-of-life management. P_{18} argued that there are no checks and balances on where the materials are going; P_{19} stated that a lot of materials that are claimed to be recycled are stockpiled, and it is not known whether they are suitable for use as no one can certify the quality; P_{24} mentioned the necessity of providing good information and awareness of PwRC; and P_{26} asked for transparency on how waste is managed in terms of dismantling, shipping, inspection, crushing, processing stockpiling conditions and quality assurance processes. Without such information, public organisations such as MainRoads WA cannot accept any PwRC to be used in construction projects. On this note, P20 suggested that sustainability rating tools developed by the Green Building Council of Australia (Green Star) and ISC can provide clarity about the lifecycle of C&D waste materials in buildings and public infrastructure projects. Doing so, they argue, would provide benefits in establishing a diversion rate, offering clarity on where waste materials go for recycling, producing high-quality materials, improving the social image of constructors and perhaps leading to economic benefits [P_{21}, P_{27}].

Market stimulation was also considered an influential factor. For P_{18}, the lack of market for C&D waste is a major push back for suppliers; P_{23} described the current conditions where a viable market for C&D waste could assist the construction industry that is struggling to obtain certain fresh construction materials (e.g., steel); and P_{24} stated that a strong market would provide a level playing field, which would also prohibit some waste operators from dodging the rules.

The economics of the waste was also raised by P_{22} as a major determinant of waste end-of-life. This participant described how the demolition operators' decisions about the waste generated during demolition are driven by the economic advantage of available waste management methods (i.e., landfilling versus recycling). A key economic factor with the strongest influence, according to P_{24}, is the landfill levy. However, the landfill levy needs to be carefully enforced to make sure that it is achieving its intended purpose. Such a levy should prohibit waste material leakage that might happen in the form of waste transfer from metropolitan areas (with higher rates) to regional ones (with no or lower rates), and illegal dumping and stockpiling [P_2]. Furthermore, from the developer's perspective [P_{25}], waste end-of-life is managed by a waste operation sub-contractor, and the only influential factors they are concerned about are the cost, labour and quality of source separation (e.g., contamination). Particularly, source separation and the cost associated was mentioned by P_{23}, who believed that divvying up the waste to minimise the risk of contamination is a determinant for effective management of waste end-of-life.

Lastly, two factors mentioned by P_{27} included setting contract conditions for better uptake of PwRC and capability building in both the human aspect (cultural change) and facilities.

Table 69. Summary of enablers and barriers specified by the participants in WA

P	Enablers	Barriers
P_{17}	• A consistent approach in testing requirements and other obligations (e.g., noise management plan) for both PwRC and virgin materials • Revisiting regulations and specifications relating to the type of PwRC application • Increase the knowledge of PwRC' properties and performance to provide confidence to buyers in buying them	• Inconsistent state government's approach in different divisions in supporting using PwRC • Unreasonably complex and testing requirements such as heavy metal concentrations for low-value applications of PwRC • Lack of resource tax on virgin materials
P_{18}	• Establishing recycling facilities closer to metropolitan areas to reduce transport cost • Seeking innovation and improve design requirements for infrastructure projects • Motivate local governments as the biggest buyers to use recycled products developing standards at the state level • Advancing the technicality of practices of application of PwRC that prevent their poor performance (e.g., implementing specific concrete with recycled content laying down technique to prevent cracking)	• Conservative approach of public organisations (e.g., MainRoads WA) in using PwRC • A tendency among stakeholders to shift the problem of waste away from them in any way possible rather than focusing on the quality and acceptance of PwRC • Lack of clarity on the destination of waste materials (e.g., where and how they recovered) claimed to be diverted from landfills by government organisations • Production of low-quality PwRC by some waste processors reduces the demand for such products • Availability of abundant cheap virgin materials in metropolitan areas in WA • The lack of appetite by public organisations (e.g., MainRoads WA) to improve design requirements of infrastructure projects to embrace new products such as PwRC • The reluctance of MainRoads WA to use PwRC as a road base • Lack of capability in the state recycling industry to provide enough PwRC required for large infrastructure projects
P_{19}	• Clarification about how decision-makers (e.g., designer, engineer) choose to use PwRC • Educating stakeholders of potential risks and benefits of using PwRC	• Limitations within government procurement policies and standard policies • Lack of knowledge of PwRC characteristics

P	Enablers	Barriers
P_{20}	• Creating a centralised platform to increase waste trading among various players • Implementing strict quality control to make sure consistent materials are procured • Fostering collaboration between industry and approving authorities to use recycled alternatives in construction projects • Educating designers who specify construction materials on PwRC alternatives	• Difficulty in using some PwRC (e.g., timber and steel) in mass production in the construction industry due to the inconsistency in size and specifications • The lack of effective implementation of carrot and stick approach in market development efforts and policies in WA • Limited usage of PwRC
P_{21}	• Developing measured regulations without an increase in costs • Providing subsidisation (e.g., Recycled Construction Product Program) as a good premise, which financially assists customers in purchasing PwRC (e.g., drainage rock and clean fill sand) for different purposes (e.g., road base) • Developing recycling centres closer to metropolitan areas to improve the economic advantage of PwRC	• Virgin material being cheaper than PwRC • Abuse in the use of government financial support for buying PwRC (e.g., paying the fund to consultant and contractor rather than end-users) • Inappropriately designed obligations for processing and using PwRC around residential developments • Unfair competition in using waste materials coming from unlicenced on-site processing and recycling facilities due to cost disparities • Lack of legal definition of waste discriminating in licence requirements for the materials going out of the location of generation for recycling versus the materials kept in the boundary of its location (licence is not required) • Government prohibition of operation of screening or crushing machinery next to residential areas due to noise and dust pollution • Risk of change in the state government policies and priorities over PwRC with change in political parties • Lack of information-sharing on C&D projects risking long-term planning for private sector • Lack of collaboration between stakeholders across waste material supply chain on the waste destination

Table 69. (Continued)

P	Enablers	Barriers
P_{22}	• Creating an online and automated waste trading portal led by government and councils where people can register their waste materials availability and location • Convincing constructors to join waste exchange platform through contract conditions • Educating the consulting engineers and architects to consider PwRC	• Lack of government effort in raising awareness or government pushing the benefits of waste recovery just like it does in carbon reduction
P_{23}	• Improving on-site source waste separation to leave the contamination out of waste mix load sent to recycling facilities • Improving technologies for optimal waste recovery	• High cost of waste management • Outdated and rigid government specifications • Lack of facilities for remanufacturing waste materials • The state government inactivity and lack of leadership in pursuing sustainability
P_{24}	• Creating material guidance (catalogue) with some clear examples (case studies) of what materials can be used and in what applications to support sustainable procurement • Providing confidence in PwRC through environmental standards to reassure buyers that these products do not contain any contaminants (e.g., heavy metals, and asbestos) • Implementing an independent audit to verify the quality of PwRC • Raising awareness and promotions of PwRC characteristics to better understand opportunities for using them • Ensuring that supply is reasonably close to where the material is needed to justify transport costs • Facilitate PwRC competition on their own merits by showing that they are completely suitable for various applications	• Lack of confidence in PwRC, which stems from many years ago • Lack of understanding about what sort of PwRC are suitable for end-users

P	Enablers	Barriers
	• Setting clear targets and requirements for preference for PwRC over virgin raw products and raw materials Leverage the requirements in contracts to get the construction sector to shift their thinking, such as giving preference to those bidders using PwRC, where possible	•
P_{25}	• Creating more sophisticated platforms for trading and buying waste materials • Creating a verification system to ensure PwRC are up to scratch and safe to go into a new building • Providing incentives for developers to motivate them to consider a certain minimum of PwRC at the planning stage • Use of sustainability rating tools by councils to drive the market for PwRC • Increase end-user (e.g., council, community) awareness of the value of using PwRC	• Lack of policy mechanisms to help drive sustainable procurement solutions • Lack of interest in paying more for PwRC that pass the cost to end-users • Lack of collaboration between constructors of assets and those who maintain it • Lack of attention to the PwRC in council's development projects • Developers' inclination to only follow what customers need, which may exclude the use of PwRC • Unavailability of information for the construction industry to judge the use of PwRC and the costs associated
P_{26}	• Targeting the extracting maximum value of PwRC through advanced waste management methods such as upcycling • Seeking long-term solutions for large volume waste streams • Supporting initiatives with minimum environmental consequences using lifecycle thinking • Use of sustainable rating tools to enable the public organisations to seek opportunities for reduced waste generation and increased uptake of PwRC	• The lack of willingness to improve processes of construction of public infrastructure projects by organisations that deliver a small number of projects within a certain period • Uncertainty about waste materials origin and handling • Promoting the interim solutions that may preclude the development of better solutions in future • The reluctance of some public organisations to seek innovations due to imposed inflexible compliance requirements and to be risk averse • Lack of recycling facilities and technologies that enable maximum possible use of waste materials value

Table 69. (Continued)

P	Enablers	Barriers
P_{27}	• Sustainable procurement and setting contract conditions in favour of using PwRC • Work with industry to increase the supply of PwRC through an optimised supply chain • Government funding to enhance recycling capability in the state • Apply sustainability rating tools (e.g., ISC) that drive right waste management practices as a benchmark for prequalification of bidders for certain public projects tendering • Sharing the ownership of sustainability rating credits with lead contractors to enable the use of PwRC	• The low economy of scale in the market to the market size and the number of players • The small market does not encourage competition and assist in improving the quality of PwRC • Strong and established commercial agreements between end-users and virgin materials suppliers • Difficulty in supply chain optimisation to accommodate new PwRC due to low competition between companies

6.3.5.6. Waste Regulations in WA

The responses to the question related to current regulations and policies revealed that not everyone is familiar with the current regulatory framework and relevant efforts to reform them (Figure 41). The majority of responses indicate the need for improving the current regulations. Only a few participants [P_{19}, P_{20}, P_{26} and P_{27}] acknowledged the current reforms at the state level to improve the effectiveness of state waste regulations. Notably, the participants with that knowledge were the ones that represent the public organisations that are involved in waste regulation and policy development. This trend clearly represents the fact that current waste policies are not being effectively communicated to stakeholders.

The responses also highlighted some of the reform priorities in current legislation. These include an emphasis on reusing waste materials and lack of extended producer responsibility policy [P_{17}]; waste end-of-life management [P_{18}]; proper distribution of landfill levy revenue [P_{19}]; encouragement of using PwRC [P_{20}]; focusing on non-hazardous materials [P_{22}]; pushing and improving the current regulations during the construction industry's busy periods [P_{23}]; clarity on definitions and requirements relating to handling waste materials as opposed to resources [P_{26}]; and legislation stimulating market development [P_{27}].

Table 70 summarises the participants' responses to the question on waste regulation in WA.

Drawing on the key literature and the findings from the interviews with participants across the C&D waste supply chain representing five Australian jurisdictions, a model to optimise the creation and stimulation of C&D waste end-markets was developed. The C&D Waste End-market Creation and Stimulation Model (Figure 42) maps how various enablers of market development can be implemented to optimise the interplay among various stakeholders in the C&D waste space. The stakeholders in this model include the government sector (public organisations and policymakers), recyclers, C&D professionals, and sustainability rating system and quality assurance auditors.

As proposed in this model, the government sector should facilitate significant improvements through two levels of strategies. At the forefront are the primary government-enabled strategies, centring on SP; raising awareness; effective recycling approvals; simple and effective material specifications;

leveraging contract conditions to use PwRC; and financial support. The secondary government-enabled improvements include: stopping illegal dumping; providing a clear definition of waste; developing supportive regulations; imposing a properly designed landfill levy; contributing to a national approach; setting targets to use PwRC; and building consistency among various public organisations in valuing (using) PwRC including encouraging local governments to purchase PwRC.

Table 70. Summary of participants' opinions on the impact of current legislation on market development in WA

P	Policy/Regulations
P_{17}	The legislation is not strong enough, and current regulations are not directed towards reducing waste. Furthermore, state waste regulations do not emphasise reuse and recycling as they should. Notably, reusing is ignored in regulations, and there is no compulsion against landfilling the resources that could be reused. The lack of extended producer responsibility is another issue that is not regulated under the current legislative framework. Currently, the decision on recycling C&D waste is primarily based on market forces, and it should be modified to be driven by regulations.
P_{18}	There is much room for improvement. For instance, there is too much emphasis on handling waste and not enough on the end product. Until the mandatory legislation that can stimulate the market are introduced, suppliers of PwRC continue to struggle with accepting their production by end-users.
P_{19}	The current legislation has definitely improved, which has resulted in a significant reduction in waste disposal. However, the issue of redistributing landfill levy revenue continues to act as a barrier in developing markets. Currently, the revenue does not go back to the implementation of waste strategies across the state.
P_{20}	The issue with the current legislation is that the landfill levy as an effective waste strategy is not accompanied by supporting the use of PwRC. There is excellent potential in WA to use PwRC as many recyclers in the state have invested in technologies and modern processes to produce high-quality materials, and the legislation should support these efforts.
P_{21}	The end-of-life waste guideline effective a few years ago was a supportive policy instrument that has been removed after a private entity sued regarding the disposal of unprocessed waste that conformed to the policy at that time. Policies that drive sustainable procurement rather than landfill levies are preferred. Furthermore, increasing the landfill levy in future legislation risks the operation of recycling facilities due to imposition of the landfill levy on residual recycling waste. Also, specifications on using PwRC in new construction materials need to capture a wider range of materials in the state.
P_{22}	The current legislation only addresses hazardous materials and focuses on regulated waste categories, and there is no legislation about nonhazardous materials such as C&D waste materials. Legislation needs to be introduced to define expectations of waste management lucidly.

P	Policy/Regulations
P_{23}	The current legislation on C&D waste, if it exists, needs to be pushed to reflect real-world conditions in the state, given a boost in the state economy and, notably, the construction industry. Such a boom in the industry, generating so much waste, and the scarcity of construction resources justify the urgent need for revisiting waste regulations. Conversely, during quiet times, there is less appetite to reform regulations required for addressing resource constraints. In revisiting the legislation, the government should send strong signals to those who oppose reforms, and the industry is accountable for communicating the current issues, such as the shortage of some construction materials that could be fixed by new legislation.
P_{24}	The *Department of Water and Environmental Regulation* has been leading an initiative called the Waste Reform program since 2019 that implements a series of reforms throughout our Environmental Protection Act. These reforms aim to make the entire material recovery system work better. This initiative attempts to look at the range of regulations out there and ensure that they are updated, more consistent, and more reliant on or supportive of our new Waste Strategy objectives. For instance, a product of these reforms is a paper called "Close the Loop," which aims to move the state towards a more sustainable economy. However, there is some work to be done.
P_{25}	The current regulations are not good enough. The current set up does not require builders and developers to source separate C&D waste and indirectly encourages mixed load waste at construction sites. The government should devise an extremely high penalty for dumping waste in landfill and builders should be strongly incentivised to reduce the amount that is going into landfills.
P_{26}	The policies about waste in their organisation are being reviewed. However, the big problem is the distinction between resource and waste, and eventually, what materials should attract the landfill levy and how waste ceases to be waste. Hence, clarity around definition and expectation is a priority in improving waste regulations. Clarification gives a good understanding of waste handling requirements and will result in minimising unsustainable waste management.
P_{27}	The state could probably get better outcomes if we had slightly more clarity on the legislation. Notably, legislation is required to stimulate better management of end-of-life management of C&D waste.

In the recycling sector, the four priority areas are to; invest in advanced processing technologies; develop infrastructure with a suitable capacity to manufacture to the scale of economy; establish such facilities close to metro areas; and generate material catalogues that detail PwRC information.

In the C&D industry, for both private and public projects, the improvements include: enabling information sharing with the recycling industry; on-site source separation; and collaborating with manufacturers to implement an extended producer responsibility policy.

Furthermore, independent auditors should be engaged to assure the quality and performance of PwRC after testing. Lastly, sustainability rating systems such as Green Star and IS credits should be leveraged to drive change in the waste management ecosystem and endorse the use of PwRC in private and public construction projects.

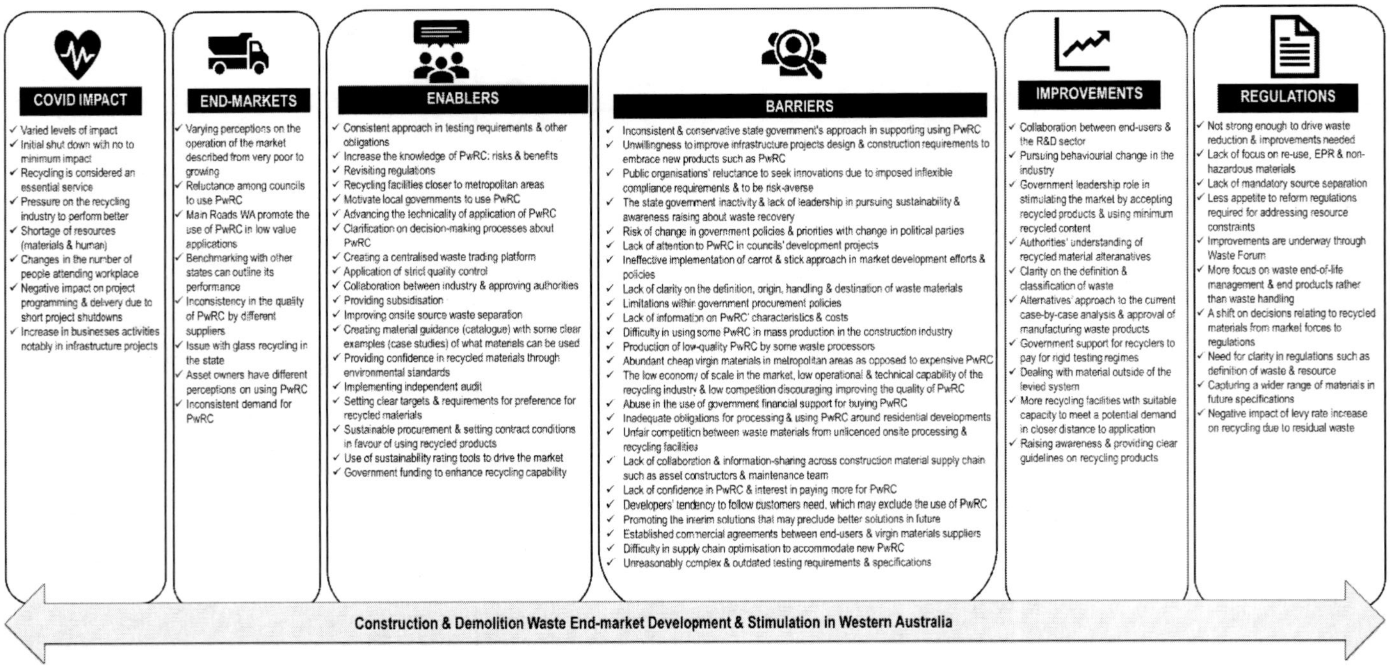

Figure 41. A conceptual model that describes C&D waste end-market characteristics in WA 6.4 Optimum model for creation and stimulation of C&D waste end-markets.

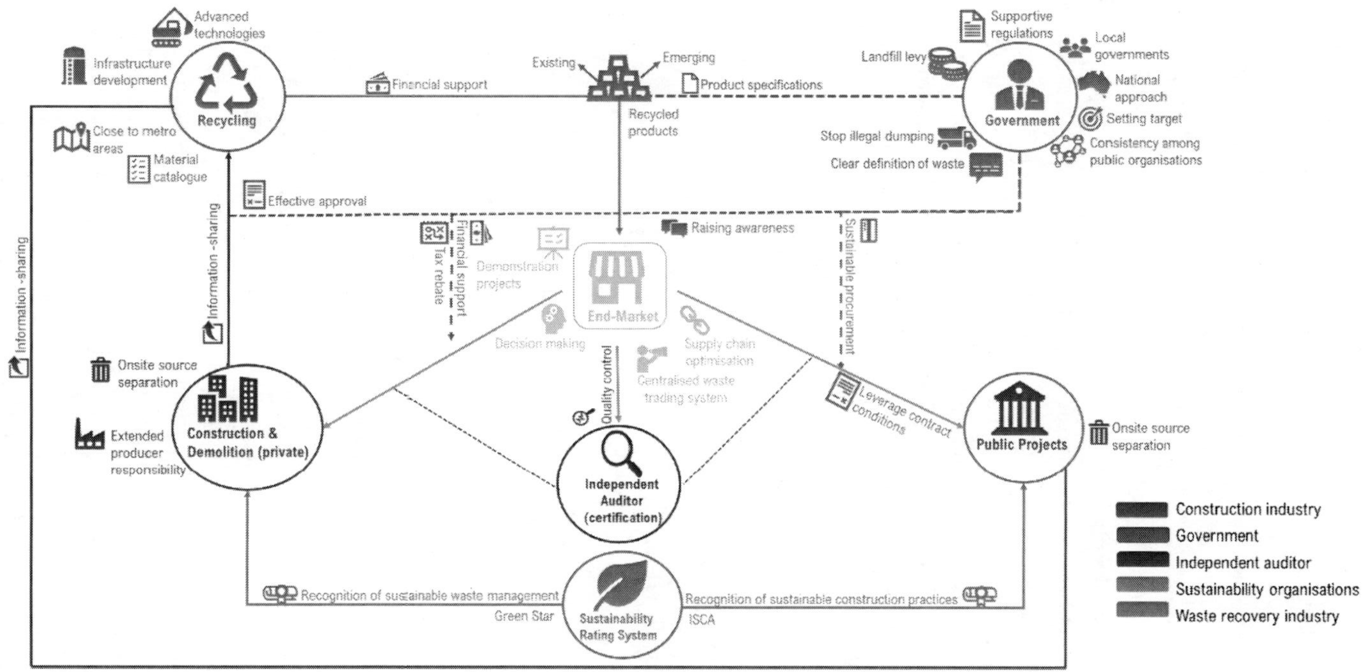

Figure 42. C&D waste end-market creation and stimulation model.

Conclusion

The creation and stimulation of end-markets for C&D waste require the involvement of various parties. Drawing on interviews with 27 C&D waste experts in five states (WA, Vic, NSW, Qld and SA), this chapter provided an insight into the main challenges and opportunities in end-market development as identified by key stakeholders.

As a major enabler, SP was frequently referred to in response to questions regarding market development and stimulation enablers. From the interviews, it can be concluded that SP has been better practised in Vic compared to NSW and SA. This trend could be attributed to the better promotion of the CE and resource efficiency in Vic. The policy support by public organisations such as *Sustainability Victoria* (Buy Recycled), *Major Road Projects Victoria* (Recycled First), and the *Department of Environment Water and Planning* (Recycling Victoria) has played a crucial role in this space. Meanwhile, in WA, initiatives such as the Roads to Reuse Program and Waste Forum can inspire other states to create such platforms to connect key stakeholders, encouraging the use of C&D waste PwRC by government organisations.

Quality assurance was identified as an important strategy to build confidence in using PwRC. In WA, the use of independent audit entities has been reported to be successful. This shows that quality assurance can occur at different stages, from the time waste is generated to when PwRC are offered to the market. Other states can learn from this success story to achieve the same results.

Overall, the participants' perceptions of the current operation of end-markets varied. It was noted that there is no tool or benchmark to precisely evaluate their performance. The participants indicated that concrete and brick have a mature market in three states (i.e., Vic, Qld and WA) and cited that glass and timber are the two waste materials with the most potential for new markets. In terms of performance, Vic has the most viable market for PwRC, which means other states and territories can follow its lead. The proximity of recycling facilities to metropolitan areas compared to virgin materials and emphasis on the use of PwRC by government agencies such as *SV* are among the chief factors that have facilitated positive outcomes.

This chapter has presented an overall picture of the key issues surrounding market development and stimulation in Australia. A suite of actionable recommendations for reforms based on comparative analysis of state-based information were provided. While the research team made every effort to recruit participants representing various stakeholder groups, it is noteworthy that the results presented in this chapter are indicative and not inclusive.

Chapter 7

Conclusion

7.1. Introduction

This book had provided a detailed evaluation of end-markets' operation and set-up for C&D waste recyclables in Australia. It examined the national and jurisdictional policy frameworks that facilitate or hinder waste exchange at state and national levels. The results demonstrate differences and similarities in the Australian jurisdictions' policy frameworks concerning the development and stimulation of end-markets for C&D waste recyclables. From this research, valuable insights have emerged into five key areas: (1) national and jurisdictional policy frameworks that affect creation and stimulation of end-markets for waste materials; (2) operation of existing end-markets for five strategic construction materials (i.e., brick, concrete, asphalt, steel and glass); (3) application of SP policies and practices at state, national and international levels; (4) impacts of the COVID-19 pandemic on the C&D waste industry; and (5) factors affecting C&D waste end-market creation and stimulation and a model for C&D waste market development. The following offers a summary of the findings regarding each of these key areas.

7.2. National and Jurisdictional Policy Frameworks

Chapter 2 presented a review of relevant policies, regulations and standards in the Australian states and territories, highlighting their unique characteristics related to development and stimulation of end-markets for C&D waste recyclables. This chapter analysed 102 federal, state and territory government policies, standards, guidelines, legislation and specifications to determine the extent to which the Australian regulatory frameworks are in favour of or against C&D waste market development. The following are the key findings from this regulation analysis:

1) Most of the primary state and territory legislation does not define C&D waste and considers waste as always waste regardless of its value or future usage;
2) Specifications of application of C&D PwRC vary between states and territories, which has a negative impact on interstate market development and stimulation;
3) In most waste strategy documents, the significance of waste market development is highlighted, and some waste strategy documents suggest recommendations to improve it; and
4) Among the states and territories, NSW and Tas have the most extensive and smallest regulatory framework relevant to C&D waste management and market development thereof, respectively.

7.3. Operation of Existing End-Markets

Chapter 3 identified major barriers to the operation of existing C&D waste end-markets. The main barriers detected included: an increase in energy and transport costs; limited knowledge of PwRC; limited technologies for waste recovery; low quality and reduced performance; lack of market availability of the PwRC; limitations caused by complex specifications and standards; limited acceptability; and negative perceptions. To overcome these barriers, several strategies were put forward, namely: increase community awareness; develop supportive regulations; promote sustainability programs; promote recycled product certifications; and advocate targeted technologies and innovations. These barriers and strategies can form a management framework that informs decisions, policies and best practice management in the AEC industry to move towards a CE.

7.4. Sustainable Procurement Practices

Chapter 4 explored how SP as a policy strategy, among other strategies, can alter the status quo of the C&D waste management system and aid the Australian AEC industry in adopting a CE. A review was conducted to assess the current condition of SP in the Australian AEC industry. The findings revealed that the two most significant obstacles are the lack of supportive organisational culture and uncertainty about PwRC quality. In addition, implementing clear and supporting regulations and maintaining transparency and good governance were identified as the two most important enablers. Furthermore, the study developed a model, The Emergent Model for Sustainable Procurement in the AEC industry, to assist stakeholders in finding opportunities for adopting SP principles at the policy formulation, education and procurement stages. To achieve this objective, a set of recommendations was provided to foster collaboration between five stakeholders (policymakers; research and education providers; end-users; recycling facility owners; and project managers in the AEC industry) to increase the likelihood of PwRC in construction project material procurement.

7.5. Impacts of COVID-19 on C&D Waste Industry

Chapter 5 presented an analysis of the responses to COVID-19-related issues collected during interviews. It showed that the study states' industries and state governments have successfully responded to the COVID-19 impacts. It was apparent that recognition of the construction and recycling industries as essential services enabled them to survive during these challenging times. The respondents provided information on how they adapted to the rapidly changing situation. Nevertheless, various levels of impact were reported by the research participants. In the recycling sector, it seems that the states of SA and NSW experienced a decline in demolition activities, resulting in lower levels of incoming waste feedstock required for sustainable operation of their facilities. However, in Vic and WA, the situation was different, and recycling facilities could receive the necessary amount of feedstock as per their capacity. These varying trends could be related to the high rate of construction activities occurring across these states at the time of the interviews necessitating waste collection and processing.

In the construction sector, respondents from the three states (i.e., Vic, SA and NSW) indicate minimal to extensive impact. Notably, in SA and Vic, the level of impact was assessed as great overall but not all AEC sectors were affected equally. The main impacts were the delay in delivery of public infrastructure projects, cost implications for implementing additional hygienic procedures and working with half capacity due to applying rotating shifts.

7.6. Optimal Model for the Creation and Stimulation of End-Markets for C&D Waste Streams

Chapter 6 uncovered a range of barriers to and enablers of market development of C&D waste recyclables; the current operation of existing end-markets; areas for improvement; and the impact of regulations and COVID-19 on such markets. Building on the information extracted from the research activities in this project, The C&D Waste End-market Creation and Stimulation Model was developed to facilitate C&D waste end-market creation and stimulation.

The proposed model establishes a roadmap showing how various AEC stakeholders should contribute to achieving the CE objectives in a collaborative manner. It assists government policymakers in developing informed policies that reduce the negative environmental impact of construction activities. Accordingly, six actionable recommendations are proposed to guide market creation and stimulation for C&D waste materials:

1) In the government sector, the primary improvements centre on SP strategies, awareness raising; adequate recycling approvals; provenance/traceability of the recycled components; simple and effective material specifications; leveraging contract conditions to use PwRC; and providing financial support.
2) The secondary government-enabled improvements include stopping illegal dumping; providing a clear definition of waste; developing supportive regulations; imposing properly designed landfill levy; contributing to a national approach; setting targets to use PwRC; and building consistency among various public organisations in valuing (using) PwRC including encouraging local governments to purchase PwRC.
3) In the recycling sector, the four priority areas are to invest in advanced processing technologies; develop infrastructure with a suitable capacity to manufacture to the scale of the economy; establish such facilities close to metro areas; and generate material catalogues that detail recycled product information including their provenance and ingredients.
4) In the AEC, for both private and public projects, the improvements include enabling information sharing with the recycling industry; on-site source separation; and collaborating with manufacturers to implement an extended producer responsibility policy.

5) Furthermore, independent auditors should be engaged to ensure PwRC' quality and performance after testing.
6) Lastly, sustainability rating systems such as Green Star and IS should be leveraged to drive change in the waste management ecosystem and endorse PwRC in private and public construction projects.

The model involves multiple facilitating strategies that theoretically contribute to a conception of the determinant factors of waste market development. Notably, the model proposes pragmatic recommendations for overcoming any challenges posed. As such, this study, and the findings of the research contained in this book as a whole, may drive circularity of waste recyclables in the housing, building and infrastructure sectors in Australia and improve the demand for C&D waste recyclables. The outcomes of this research may assist policymakers to identify the opportunities and challenges in sustainable waste management; encourage the industry to reduce its impact on the environment; motivate waste recovery facilities to find demand for their products; and help the public to benefit from a sustainable society.

References

[1] Aslam, M. S., Huang, B. and Cui, L. 2020. Review of construction and demolition waste management in China and USA. *Journal of Environmental Management*, 264, 110445.

[2] Blue Environment. *National waste report 2020*. Retrieved from https://bit.ly/3pnRE5D.

[3] Shooshtarian, S., Maqsood, T., Wong, P. S., Yang, R. J. and Khalfan, M. 2020. Review of waste strategy documents in Australia: Analysis of strategies for construction and demolition waste. *International Journal of Environmental Technology and Management* 23(1): 1-21.

[4] Ellen MacArthur Foundation. 2019. *What is a circular economy?* Retrieved from https://bit.ly/3p8KhRe.

[5] Halog, A., Balanay, R., Anieke, S. and Yu, T. Y. 2021. Circular economy across Australia: Taking stock of progress and lessons. *Circular Economy and Sustainability*. 1 (2021): 283–301.

[6] NSW Government. 2019. *NSW circular economy policy statement too good to waste*. 2019. Retrieved from https://bit.ly/3zQxQ1X.

[7] SA Government. 2020. *South Australia's Waste Strategy 2020-2025: Supporting the Circular Economy*. Adelaide, Australia 2020. Retrieved from https://bit.ly/41kEPMc.

[8] Parry-Husbands, H., Zwart, J., Rice-Jenkins, B., Gilling, R. and Garofano, N. 2021. *Circularity in Australian Business: Awareness, Knowledge and Perceptions*. AEC Hub (Planet Ark Environmental Foundation) (Sydney, Australia 2021). Retrieved from https://bit.ly/31vAdJJ.

[9] Mignacca, B. and Locatelli, G. 2021. Modular circular economy in energy infrastructure projects: Enabling factors and barriers. *Journal of Management in Engineering*. 37(5):04021053.

[10] Melles, G. 2021. Figuring the transition from circular economy to circular society in Australia. *Sustainability* 13(19): 10601.

[11] Shooshtarian, S., Khalfan, M., Maqsood, T., Wong, P. S. and Yang, R. J. 2020. Market development for construction and demolition waste stream in Australia. *Journal of Construction Engineering, Management & Innovation*. 3(3): 220-231.

[12] Oyedele, L. O., Ajayi, S. O. and Kadiri, K. O. 2014. Use of recycled products in UK construction industry: An empirical investigation into critical impediments and strategies for improvement. *Resources, Conservation and Recycling*. 93:23-31.

[13] Ding, Z., Nie, W. and Wu, H. 2021. Investigating the connection between stakeholders' purchase intention and perceived value of construction and demolition waste recycled products. *Environment, Development and Sustainability.* 2021:1-19.

[14] Jain M. S. 2021. A mini review on generation, handling, and initiatives to tackle construction and demolition waste in India. *Environmental Technology & Innovation.* 22:101490.

[15] Bougrain, F. 2017. Impact of green public procurement on the market of recycled concrete. *HISER International Conference: Advances in Recycling and Management of Construction and Demolition Waste.* Delft, The Netherlands: Delft University of Technology.

[16] Nunes K. and Mahler C. 2020. Comparison of construction and demolition waste management between Brazil, European Union and USA. *Waste Management & Research.* 38(4):415-422.

[17] Caldera, S., Ryley, T. and Zatyko, N. 2020. Enablers and barriers for creating a marketplace for construction and demolition waste: A systematic literature review. *Sustainability.* 12(23), p.9931.

[18] Solís-Guzmán, J., Marrero, M., Montes-Delgado, M. V. and Ramírez-De-Arellano, A. 2009. A Spanish model for quantification and management of construction waste. *Waste Management*, 29, 2542-2548.

[19] Menegaki, M. and Damigos, D. 2018. A review on current situation and challenges of construction and demolition waste management. *Current Opinion in Green and Sustainable Chemistry*, 13, 8-15.

[20] Zheng, L., Wu, H., Zhang, H., Duan, H., Wang, J., Jiang, W., Dong, B., Liu, G., Zuo, J. and Song, Q. 2017. Characterizing the generation and flows of construction and demolition waste in China. *Construction and Building Materials*, 136, 405-413.

[21] US EPA 2019. *Advancing sustainable materials management: 2017 Fact sheet.* EPA 530-F-19-007. United States Environmental Protection Agency. Retrieved from https://bit.ly/43z6WJi.

[22] Peng, C. L., Scorpio, D. E. and Kibert, C. J., 1997. Strategies for successful construction and demolition waste recycling operations. *Construction Management & Economics*, 15(1), 49-58.

[23] National Waste Policy. 2018. Less Waste. More Resources. In: *Department of Agriculture*, W. A. T. E. (ed.). Canberra, Australia. Retrieved from https://bit.ly/3KYn8wu.

[24] *Environment and Communications References Committee*. 2018. Never waste a crisis: the waste and recycling industry in Australia. Canberra, Australia Parliament of Australia. Retrieved from https://bit.ly/3KEtgc9.

[25] Australia Government. 2018. Product Stewardship Act 2011. Retrieved from https://bit.ly/3ohYcra.

[26] Austroads. 2022. *Guide to Pavement Technology Part 4E: Recycled Materials*. Retrieved from https://bit.ly/409Yn54.

[27] Development control code for best practice waste management in the ACT 2019. Transport Canberra and City Services Directorate. Retrieved from https://bit.ly/3ddEnIr.

[28] ACT Waste-to-Energy Policy 2020-25- ACT Government. Retrieved from https://bit.ly/3IgXGyP.

[29] NSW EPA. 2015. *NSW Energy from Waste Policy Statement* (updated in 2021). Retrieved from https://bit.ly/3KZGcKX.

[30] ARRB Group Ltd. 2019. *Specifications for recycled crushed glass as an engineering material*. Retrieved from https://bit.ly/3zZlOmL.

[31] *Queensland Procurement Policy 2019*. Department of Housing and Public Works. Retrieved from https://bit.ly/3lsN5HC.

[32] Green Industries SA Act 2004. Retrieved from https://bit.ly/3pncLX7.

[33] ABS. 2020. *Waste account, Australia, experimental estimates*. Retrieved from https://bit.ly/3ppc7r1.

[34] ABS. 2020. *Building activity, Australia*. Retrieved from https://bit.ly/3pdAokx.

[35] Australian Government. 2019. *National waste policy action plan*. Retrieved from https://bit.ly/2TtBTiy.

[36] Blue Environment. 2020. *National waste database*. Retrieved from https://bit.ly/3zIju3l.

[37] Wu, H., Zuo, J., Yuan, H., Zillante, G. and Wang, J. 2020. Cross-regional mobility of construction and demolition waste in Australia: An exploratory study. *Resources, Conservation and Recycling*, 156, 104710.

[38] *Waste Management Review*. 2018. NSW crackdown targets illegal waste transporters. Retrieved from https://bit.ly/3vYbB6k.

[39] ACT NoWaste. 2018. *Waste feasibility study roadmap and recommendations*. Prepared for the ACT Government. Retrieved from https://bit.ly/41LNZSj.

[40] Planet ARK. 2020. *Business Recycling*. Retrieved from https://businessrecycling.com.au/.

[41] Canberra Business Chamber 2014. *Building and construction waste materials: Reduce, Reuse and Recycle-Opportunities and strategies for the Capital region*. Canberra Sustainability Special Interest Group. Retrieved from https://bit.ly/3A08TkK.

[42] ACT Government. 2020. ACT waste-to-energy policy 2020-25. Retrieved from https://bit.ly/3MtGU4i.

[43] *Canberra Concrete Recyclers*. 2020. Retrieved from https://bit.ly/43g6D66.

[44] *Canberra Construction Recyclers*. 2020. Retrieved from https://bit.ly/3UGpzHl.

[45] ACT recycling. 2020. Retrieved from https://bit.ly/3L7byiZ.

[46] Riqtact. 2019. *Canberra roads to be paved with recycled plastic as Government trials new type of asphalt*. Retrieved from https://bit.ly/2TqxhJ6.

[47] Australian Government. 2020. *Joint Media Release: $21 million for better recycling for the ACT*. Retrieved from https://bit.ly/3ieOelg.

[48] *Waste Management Review*. 2018. NSW Crackdown Targets Illegal Waste Transporters. Retrieved from https://bit.ly/3vYbB6k.

[49] Blue Environment. 2019. *National waste data and reporting cycle 2017-19. State and territory feedback and suggested improvements*. Retrieved from https://bit.ly/2S4cbAZ.

[50] *Population Australia*. 2020. Population of New South Wales 2021. Retrieved from https://bit.ly/3KpXak9.

[51] *NSW Industry Report*. 2020. Infrastructure and construction. Retrieved from https://bit.ly/3sHEGC4.

[52] PWC. 2019. *NSW Waste Sector Volume II: Situational Analysis*. Retrieved from https://bit.ly/3bRy0eF.

[53] Local Government NSW. 2020. *Recycled materials in roads and pavements a technical review*. Retrieved from https://bit.ly/35VCZXW.

[54] *The Recyclers*. 2020. Retrieved from https://bit.ly/3oaYKij.

[55] *Current scrap metal prices Australia*. 2021. Scrap metal prices Sydney. Retrieved from https://bit.ly/2MnsEx2.

[56] *Shop crushed glass aggregate*. Retrieved from https://bit.ly/2Mhpyul.

[57] NSW EPA. 2016. *New minimum standards for managing construction and demolition waste in NSW*. Retrieved from https://bit.ly/39LELMq.

[58] GHD. 2009. *Packaging Stewardship Forum*, Australian Food and Grocery Council. Retrieved from https://bit.ly/35UZcp5.

[59] Edge Environment Pty Ltd. 2012. *Construction and demolition waste guide recycling and re-use across the supply chain*. Retrieved from https://bit.ly/2Kuruz4 f.

[60] Cheng, A. 2020. Council trials recycled glass roads. *Government News*. Retrieved from https://bit.ly/3MLJh2A.

[61] Laclette, A. 2021. *Circular economy projects eligible for grants*. Planet Ark. Retrieved from https://bit.ly/3bRaBtL.

[62] NSW EPA. 2021. *Civil construction market program grants*. Retrieved from https://bit.ly/38WHdjV.

[63] Department of Treasury and Finance. 2020. *Northern Territory economy*. Retrieved from https://bit.ly/3bRa0Z3.

[64] City of Darwin. 2020. *Overview - Shoal Bay Waste Management Facility*. Retrieved from https://bit.ly/38XYzwB.

[65] Mathur, D., O'Leary, R. and Gerritsen, R. 2016. *Reducing building waste in Alice Springs*. Charles Darwin University. Retrieved from https://bit.ly/3KGkIkM.

[66] City of Darwin. 2020. *Overview - Shoal Bay Waste Management Facility*. Retrieved from https://bit.ly/38XYzwB.

[67] Mirage. 2020. *Recycled asphalt used for the first time in NT*. Retrieved from https://bit.ly/3j7lb1d

[68] Australian Capital Territory Government. 2018. Waste and recycling industry in Australia- Submission 09: NT Government's response (*Environment and Communications References Committee*). Retrieved from https://bit.ly/3MFx XVC.

[69] Population Australia. 2023. *Population of Queensland 2023*. Retrieved from https://bit.ly/39LqxLj.

[70] Queensland Government. 2019. *Recycling and waste in Queensland*. Retrieved from https://bit.ly/38Zhojl.

[71] Jones, L. 2019. Roads Online: *Queensland achieves first recycled plastic road*. Retrieved from https://bit.ly/39K0jJm.

[72] Queensland Government. 2018. Discussion paper: *Transforming Queensland's Recycling and Waste Industry*. Retrieved from https://bit.ly/3oWflBY.

[73] Baird, L. 2020. Cleanaway can negotiate the rocky regulatory road. *Financial Review*. Retrieved from https://bit.ly/3sEbImu.

[74] Australian Landfill Owners Association. 2018. *Landfill levy for Queensland*, A submission to the Department of Environment and Science. Retrieved from https://bit.ly/2Nn30sL.

[75] Moreton Bay Recycling. 2019. *Concrete recycling in Brisbane: Tip vs specialist recycling centre*. Retrieved from https://bit.ly/2M9yJNb.

[76] Jones, L. 2019. Roads Online: *Queensland achieves first recycled plastic road*. Retrieved from https://bit.ly/39K0jJm.

[77] Alex Fraser. 2020. *Bruce Highway upgrade program*. Retrieved from https://bit.ly/4197YdJ.

[78] Alex Fraser. 2018. *Clem Jones Tunnel – Project Profile*. Retrieved from https://bit.ly/3Kx6pPy.

[79] *Waste Management Review*. 2018. Queensland opens $100M funding program for waste and recycling. Retrieved from https://bit.ly/39M2jki.

[80] ABS. 2020. *National, state and territory population*. Retrieved from https://bit.ly/3o1599S.

[81] McCaskill, M. 2020. *Britannica: South Australia*. Retrieved from https://bit.ly/3p7hF9y.

[82] Ai Group. 2020. *Economic outlook: South Australia*. Retrieved from https://bit.ly/38WdwPT.

[83] WT Partnership. 2019. *Australian construction market conditions report*. Retrieved from https://bit.ly/2M4maTu.

[84] Government of SA. *SA leads nation in building and construction jobs growth*. Retrieved from https://bit.ly/3bPKgMG.

[85] Green Industries SA. 2015. *South Australia's waste strategy 2015-2020*. Retrieved from https://bit.ly/3qFRt6f.

[86] Green Industries SA.2020. *South Australia's recycling activity survey 2018-19 report*. Retrieved from https://bit.ly/3qBQQdP.

[87] Green Industries SA. 2020. *Circular economy in action in south Australia: Pre-cycle – making building sites cleaner, greener, and more efficient*.

[88] The Lead. 2018. *Pre-cycling in construction industry saves money and environment*. Retrieved from https://bit.ly/3p0ieBX.

[89] Sustainable Aggregates SA. 2018. *Cooper Street Kudla*. Retrieved from https://bit.ly/3pykT5W.

[90] Realestate. 2021.

[91] GISA. 2021. *Circular economy market development grants*. Retrieved from https://bit.ly/36uviln.

[92] GISA. 2021. *Recycling modernisation grant program*. Retrieved from https://bit.ly/2MJxMLO.

[93] GISA. 2020. *A vision for a circular economy: Waste Strategy 2020-2025*. Retrieved from https://bit.ly/3j7e6xF.

[94] Scott, P. 2017. *Britannica: Tasmania Island and state, Australia*. Retrieved from https://bit.ly/3sQg0Ya.

[95] Inside Construction. 2019. *Skills and construction boom in Tasmania*. Retrieved from https://bit.ly/3qvTIss.

[96] Department of Primary Industries, Parks, Water and Environment. 2019. Draft *Waste Action Plan*. Retrieved from https://bit.ly/39LLjKN.

[97] Urban EP. 2019. *Feasibility study into a statewide waste management arrangement*. Retrieved from https://bit.ly/3bVp9Zq.

[98] Crerar, C. 2019. *Renew: Inside the war on construction waste*. Sanctuary. Retrieved from https://bit.ly/3A2xFAI.

[99] Resource Work Cooperative. 2020. Retrieved from https://bit.ly/2M5SkOq.

[100] Kingsborough Council. 2020. *Recycled roads in Kingston*. Retrieved from https://bit.ly/39l5RuE.

[101] Downer Group. 2020. *Reconophalt*. Retrieved from https://bit.ly/3KGsgE5.

[102] Northern Tasmanian Waste Management Group. 2018. *Five-year strategy: 2017–2022*. Retrieved from https://bit.ly/3KWJ4Iw.

[103] Population of Australia. 2023. *Population of Victoria 2023*. Retrieved from https://bit.ly/3ojnLCt.

[104] Live in Melbourne. 2021. *Transport, defence and construction technologies*. Retrieved from https://bit.ly/3pnYPej.

[105] Sustainability Victoria. 2014. *Market summary – recycled brick, stone and concrete*. Retrieved from https://bit.ly/3poOSNL.

[106] Sustainability Victoria. 2018. *State-wide waste and resource recovery infrastructure plan (Victoria 2017)*. Retrieved from https://bit.ly/3ae0iNI.

[107] Infrastructure Victoria. 2020. *Advice on recycling and resource recovery infrastructure*. Retrieved from https://bit.ly/3an2OBB.

[108] Waste Management Review. 2019. *Alex Fraser opens high recycled technology asphalt plant*. Retrieved from https://bit.ly/2NLV2JS.

[109] Department of Environment, Land, Water and Planning. 2020. *Recycling Victoria A new economy*. Retrieved from https://bit.ly/3j4pEBV.

[110] Bingo Industries. 2021. *Aggregate* Retrieved from https://bit.ly/3iRXI4f.

[111] Recycled Brick Centre. 2020. Retrieved from https://bit.ly/3abqj07.

[112] Newcomb. 2021. *Recycled asphalt*. Retrieved from https://bit.ly/3oslDbx.

[113] Current scrap metal prices Australia. 2021. *Scrap metal prices Perth*. Retrieved from https://bit.ly/3678WMU.

[114] Sustainability Victoria. 2019. *Guide to better practice at resource recovery centres*. Retrieved from https://bit.ly/3pvyv1Q.

[115] Victorian Government. 2016. *Resource recovery infrastructure and landfills in Melbourne*. Retrieved from https://bit.ly/2MK1YH2.

[116] A circular economy for Victoria. 2019. *Sustainability Victoria*. Retrieved from https://bit.ly/3t3KyFE.

[117] Alex Fraser. 2018. *Dingley Bypass*. Retrieved from https://bit.ly/3a7s7r1.

[118] InsideConstruction. 2020. *Grants available for construction projects using recycled materials*. Retrieved from https://bit.ly/38WcFyF.

[119] Population Australia. 2023. *Population of Western Australia 2023*. Retrieved from https://bit.ly/36e0dZB.

[120] WA Department of Jobs, Tourism, Science and Innovation. 2020. *Western Australia economic profile*. Retrieved from https://bit.ly/39Rwvdx.

[121] *ABC News*. 2018. Subcontractors are the human face of a construction industry in crisis. https://ab.co/2Lxvyea.

[122] Waste Authority. 2020. *Roads to Reuse*. Retrieved from https://bit.ly/391jhvz.

[123] Waste Authority. 2020. *Recycling activity in Western Australia 2018-19*. Retrieved from https://bit.ly/2MrxMQE.

[124] WA Department of Jobs, Tourism, Science and Innovation. 2020. *Western Australia economic profile*. Retrieved from https://bit.ly/39Rwvdx.

[125] Active Sustainability. 2020. *WA construction resources - recovered construction and demolition materials resource guide*. Retrieved from https://bit.ly/36a9Po1.

[126] REDe Recycled Bricks. *2021 products*. Retrieved from https://bit.ly/36cvNXz.

[127] Current scrap metal prices Australia. 2021. *Scrap metal prices Perth*. Retrieved from https://bit.ly/3678WMU.

[128] Smithers. 2019. *Biggest recycling plant in southern hemisphere opens in WA*. The West Australian. Retrieved from https://bit.ly/2YdUxKE.

[129] WA Government. 2020. *Recycle Right*. Retrieved from https://bit.ly/39g5Ic1.

[130] Waste Authority. 2019. *Roads to reuse pilot project case study*. Retrieved from https://bit.ly/2M5IWun.

[131] Building for Tomorrow. 2019. *Russell Road to Roe Highway*. Retrieved from https://bit.ly/41zXQdP.

[132] Jones, L. 2019. *Boral uses for recycled materials in Perth Street*. Roads and Infrastructure Australia. Retrieved from https://bit.ly/3ooeCbw.

[133] Quary.2020. *Suburban roads paved with sustainable asphalt*. Retrieved from https://bit.ly/3iSiAYM.

[134] Boral. 2020. *INNOVO™*. Retrieved from https://bit.ly/3UEc6Qk.

[135] Department of Water and Environmental Regulation. 2018. *Submission: Waste levy and waste management: proposed approaches for legislative reform*. Retrieved from https://bit.ly/2Nq3Che.

[136] Umar, U. A., Shafiq, N., Malakahmad, A., Nuruddin, M. F. and Khamidi, M. F. 2017. A review on adoption of novel techniques in construction waste management and policy. *Journal of Material Cycles and Waste Management.* 19, 1361–1373.

[137] Tam, V. W. 2009. Comparing the implementation of concrete recycling in the Australian and Japanese construction industries. *Journal of Cleaner Production.* 17, 688–7029.

[138] Dunant, C. F., Drewniok, M. P., Sansom, M., Corbey, S., Allwood, J. M. and Cullen, J. M. 2017. Real and perceived barriers to steel reuse across the UK construction value chain. *Resources Conservation and Recycling*. 126, 118–131.

[139] Colin, R. 2019. *Systems for reuse, repurposing and upcycling of existing building components*. PhD Thesis, University College London, London, UK.

[140] Hiete, M., Stengel, J., Ludwig, J. and Schultmann, F. Matching construction and demolition waste supply to recycling demand: A regional management chain model. *Building Research and Information*. 2011, 39, 333–351.

[141] Wang, J., Li, Z. and Tam, V. W. 2014. Critical factors in effective construction waste minimization at the design stage: A Shenzhen case study, China. *Resources Conservation and Recycling*. 82, 1–7.

[142] Begum, R. A., Satari, S. K. and Pereira, J. J. 2010. Waste generation and recycling: comparison of conventional and industrialized building systems. *American Journal of Environmental Science*. 6, 383–388.

[143] Denyer, D. and Tranfield, D. 2009. Producing a systematic review. *The Sage handbook of organizational research methods*. Buchanan, D. A., Bryman, A., Eds. Sage Publications: Thousand Oaks, CA, USA.

[144] Sustainability Victoria. 2014. *Market summary – recycled brick, stone and concrete*. Retrieved from https://bit.ly/3poOSNL.

[145] Liu, J., Nie, J. and Yuan, H. 2020. Interactive decisions of the waste producer and the recycler in construction waste recycling. *Journal of Cleaner Production*. 256, 120403.

[146] Bolden, J., Abu-Lebdeh, T. and Fini, E. 2013. Utilization of Recycled and Waste Materials in Various Construction Applications. *American Journal of Environmental Science*. 9, 14–24.

[147] He, L. and Yuan, H. 2020. Investigation of construction waste recycling decisions by considering consumers' quality perceptions. *Journal of Cleaner Production*. 259, 120928.

[148] Sánchez, I. G. and Lauritzen, E. K. 2004. Integrated decontamination and rehabilitation of buildings, structures and materials in urban renewal-a European project for a sustainable city concept. In *Proceedings of the International RILEM Conference on the Use of Recycled Materials in Buildings and Structures*, Barcelona, Spain, 8–11 November 2004.

[149] Henry, M and; Kato, Y. 2012. Perspectives on Sustainable Practice and Materials in the Japanese Concrete Industry. *Journal of Material Civil Engineering*. 24, 275–288.

[150] Knoeri, C., Binder, C. R. and Althaus, H.-J. 2011. Decisions on recycling: Construction stakeholders' decisions regarding recycled mineral construction materials. *Resources, Conservation and Recycling*. 55, 1039–1050.

[151] Ghaffar, S. H., Burman, M. and Braimah, N. 2020. Pathways to circular construction: An integrated management of construction and demolition waste for resource recovery. *Journal of Cleaner Production*. 244, 118710.

[152] Eisenstein, W., Fuertes, G., Kaam, S., Seigel, K., Arens, E. and Mozingo, L. 2017. Climate Co-Benefits of Green Building Standards: Water, Waste and Transportation. *Building Research and Information*. 45, 828–844.

[153] Shooshtarian, S., Maqsood, T., Wong, S. P., Malik, K. and Yang, R. 2019. Green construction and construction and demolition waste management in Australia. In *Proceedings of the 43rd AUBEA Conference: Built to Thrive: Creating Buildings and Cities That Support Individual Well-Being and Community Prosperity,* Noosa, Australia, 6–8 November 2019.

[154] Lu, W., Chi, B., Bao, Z. and Zetkulic, A. 2019. Evaluating the effects of green building on construction waste management: A comparative study of three green building rating systems. *Building and Environment.* 155, 247-256.

[155] Leipold, S. and Petit-Boix, A. The circular economy and the bio-based sector—Perspectives of European and German stakeholders. *Journal of Cleaner Production.* 2018, 201, 1125–1137.

[156] Tam, V. W., Kotrayothar, D. and Loo, Y.-C. 2009. On the prevailing construction waste recycling practices: A Southeast Queensland study. *Waste Management and Research.* 27, 167–174.

[157] Shooshtarian, S., Maqsood, T., Caldera, S. and Ryley, T. 2022. Transformation towards a circular economy in the Australian Construction and Demolition Waste management System. *Sustainable Production and Consumption*, 30, 89-106.

[158] Davis, P., Aziz, F., Newaz, M. T., Sher, W. and Simon, L. 2021. The classification of construction waste material using a deep convolutional neural network. *Automation in Construction.* 122, 103481.

[159] Alhola, L., Ryding, S. O., Salmenperu, H. and Busch, N. J. 2019. Exploiting the potential of public procurement: Opportunities for circular economy. *Journal of Industrial Ecology.* 23(1): 96-109.

[160] Brammer, S. and Walker, H. 2011. Sustainable procurement in the public sector: an international comparative study. *International Journal of Operations & Production Management.* 31 (4): 452-476.

[161] Pagell, M. and Shevchenko, A. 2014. Why research in sustainable supply chain management should have no future. *Journal of Supply Chain Management.* 50(1):44-55.

[162] Diófási, O. and Valkó, L. 2014. Step by step towards mandatory green public procurement. *Periodica Polytechnica Social and Management Sciences.* 22(1):21-27.

[163] PIANOo. 2020. *Circular procurement.* Retrieved from https://bit.ly/3mDtmIK.

[164] Van Oppen, C., Croon, G. and De Vroe, D. B. 2018. *Circular procurement in 8 steps*, Netherland, Dutch Ministry of Infrastructure and Water Management. Retrieved from https://bit.ly/40dPl6W.

[165] Australian Government. 2003. *Environmental purchasing guide: An Australian government initiative.* The Department of the Environment and Heritage. Retrieved from https://bit.ly/3KBN7IT.

[166] NCDENR. 2010. *What is environmentally preferable procurement?* North Carolina Department of Environment and Natural Resources.

[167] Pick, F. 2017. *Waste as a resource: tools for construction and demolition waste management.* Metro Vancouver in conjunction with the National Zero Waste Council. Retrieved from https://bit.ly/2Ob6dvO.

[168] Craighill, A. and Powell, J. C. 1999. *A lifecycle assessment and evaluation of construction and demolition waste.* Retrieved from https://bit.ly/3mqgJRm.

[169] UNEP. 2019. *Enhancing the uptake and impact of corporate sustainability reporting.* Retrieved from https://bit.ly/2RyZoX1.

[170] Da costa, B. B. and Da Motta, A. L. T. 2019. Key factors hindering sustainable procurement in the Brazilian public sector: a Delphi study. *International Journal of Sustainable Development and Planning*, 14, 152-171.

[171] US EPA (Agency USEP). 2021. *Sustainable marketplace: Greener products and services*. Retrieved from https://bit.ly/34nXVFQ.

[172] Geng, Y. and Doberstein, B. 2008. Greening government procurement in developing countries: Building capacity in China. *Journal of Environmental Management*, 88, 932-938.

[173] Wang C. H., Qiao Y. H. and Li, W. G. 2011. On challenges of policy implementation in China green public procurement (EBM). *Proceedings of the 2 International Conference on Engineering and Business Management*. Scientific Research Publishing Inc. 22-24 March 1064–1068.

[174] Tan, Y. Shen, L. and Yao, H. 2011. Sustainable construction practice and contractors' competitiveness: A preliminary study. *Habitat International*. 35(2): 225-230.

[175] Aboginije, A., Aigbavboa, C. Thwala, W. and Samuel, S. 2020. Determining the impact of construction and demolition waste reduction practices on green building projects in Gauteng Province, South Africa. *Proceedings of the 10 Annual International Conference on Industrial Engineering and Operations Management*, 10–12 March 2428–2437.

[176] Nicklaus, D. and Kochert, T. 2015. *Country profile —France: 2015 review of material resource efficiency policies in Europe*. European Environment Agency.

[177] Competence Centre for Sustainable and Innovative Public Procurement. 2019. *Sustainability of public procurement in Finland*. Retrieved from https://bit.ly/3bPFcYk.

[178] REEVE Consulting. 2011. *The 2011 Report on the state of municipal sustainable procurement in Canad*a. Retrieved from https://bit.ly/409JAY1.

[179] Ruparathna, R. and Hewage, K. 2015. Sustainable procurement in the Canadian construction industry: challenges and benefits. *Canadian Journal of Civil Engineering*. 42(6):417-426.

[180] Bohari, A. A. M., Skitmore, M., Xia, B. and Teo, M. 2017. Green oriented procurement for building projects: Preliminary findings from Malaysia. *Journal of Cleaner Production*, 148, 690-700.

[181] He, Q., Shi, S. and Hu, M. 2014. *Exploring a long-term mechanism of construction and demolition waste recycling: A case of Chongqing*. Springer Berlin Heidelberg.

[182] Lingegård, S., Olsson, J. A., Kadefors, A. and Uppenberg, S. 2021. Sustainable public procurement in large infrastructure projects—policy implementation for carbon emission reductions. *Sustainability*, 13, 11182.

[183] Caldera, S., Mohamed, S. and Feng, Y. 2022. Evaluating the COVID-19 impacts on sustainable procurement: experiences from the Australian built environment sector. *Sustainability*, 14, 4163.

[184] Ershadi, M., Jefferies, M., Davis, P. and Mojtahedi, M. 2021. Achieving sustainable procurement in construction projects: The pivotal role of a project management office. *Construction Economics and Building*, 21, 45-64.

[185] Shooshtarian, S., Khalfan, M., Maqsood, T., Wong, P. S. and Yang, R. J. 2020. Market development for construction and demolition waste stream in Australia. *Journal of Construction Engineering, Management & Innovation*, 3(3): 220-231.

[186] Sanchez, A. X., Lehtiranta, L., Hampson, K. D. and Kenley, R. 2014. Evaluation framework for green procurement in road construction. *Smart and Sustainable Built Environment*. 3(2): 153-169.

[187] Udawatta, N., Zuo, J., Chiveralls, K. and Zillante, G. 2015. Improving waste management in construction projects: An Australian study. *Resources, Conservation and Recycling*, 101, 73-83.

[188] Park, J. and Tucker, R., 2017. Overcoming barriers to the reuse of construction waste material in Australia: a review of the literature. *International Journal of Construction Management*, 17(3), 228-237.

[189] Shooshtarian, S., Caldera, S., Maqsood, T., Ryley, T. and Khalfan, M. 2021. An investigation into challenges and opportunities in the Australian construction and demolition waste management system. *Engineering, Construction and Architectural Management*. 29 (10): 4313-4330.

[190] Shooshtarian, S., Hosseini, M. R., Kocaturk, T., Ashraf, M., Arnel, T. and Doerfler, J. 2021. *The circular economy in the Australian built environment: The state of play and a research agenda Deakin University.*

[191] Fellows, R. F. and Liu, A. M. 2021. *Research methods for construction*, John Wiley & Sons.

[192] Axinn, W. G. & Pearce, L. D. 2006. *Mixed method data collection strategies*, Cambridge University Press.

[193] Hyder Consulting Pty Ltd. 2011. *Construction and demolition waste status report: Management of construction and demolition waste in Australia*. Department of Energy and Environment. Retrieved from https://bit.ly/3KvrKZH.

[194] Davis, P., Simon, L., Sher, W., Tang, P. and Newaz, M. T. 2019. Key solutions for construction and demolition (C&D) waste management in NSW, Australia. 43rd *AUBEA Conference: Built to Thrive: Creating Buildings and Cities That Support Individual Well-Being and Community Prosperity*. Noosa, Australia: CQ University.

[195] Australian Government 2020. Sustainable procurement guide: A practical guide for commonwealth entities. In: *Department of Agriculture*, W. A. T. E. (ed.). Canberra, Australia November. CC BY 4.0.

[196] *Australian ISO 20400 Committee*. 2018. Building the business case for sustainable procurement in Australia – guidance. Retrieved from https://bit.ly/ 3ybOcm0.

[197] Local Government NSW. 2021. *Sustainable Procurement Guide: for local government in NSW*. Retrieved from https://bit.ly/414oxay.

[198] Delmonico, D., Jabbour. C. J. C., Pereira, S. C. F., de Sousa Jabbour, A. B. L., Renwick, D. W. S. and Thomé, A. M. T. 2018. Unveiling barriers to sustainable public procurement in emerging economies: Evidence from a leading sustainable supply chain initiative in Latin America. *Resources, Conservation and Recycling*, 134:70-79.

[199] McMurray, A. J., Islam, M. M., Siwar, C. and Fine, J. 2014. Sustainable procurement in Malaysian organizations: Practices, barriers and opportunities', *Journal of Purchasing and Supply Management.* 20(3):195-207.

[200] Islam, M. M., Murad, M. W., McMurray, A. J. and Abalala, T. S. 2017. Aspects of sustainable procurement practices by public and private organisations in Saudi Arabia: an empirical study. *International Journal of Sustainable Development & World Ecology.* 24(4):289-303.

[201] Adell, A. and Schaefe, B. 2019. *Green public procurement in the Republic of Korea: A decade of progress and lessons learned.* KEITI. Retrieved from https://bit.ly/3KYkeYR.

[202] United Nations Environmental Programme. 2016. *Measuring and communicating the benefits of sustainable public procurement.* Retrieved from https://bit.ly/3A1G5rL.

[203] Shooshtarian, S., Caldera, S., Maqsood, T. and Ryley, T., 2022. Evaluating the COVID-19 impacts on the construction and demolition waste management and resource recovery industry: experience from the Australian built environment sector. *Clean Technologies and Environmental Policy.* 24(10), 3199-3212.

[204] Zhu, Q., Geng, Y. and Sarkis, J. 2013. Motivating green public procurement in China: An individual level perspective. *Journal of Environmental Management.* 126, 85-95.

[205] Sajjad, A., Eweje, G. and Tappin, D. 2020. Managerial perspectives on drivers for and barriers to sustainable supply chain management implementation: Evidence from New Zealand. *Business Strategy and the Environment*, 29, 592-604.

[206] Ogunsanya, O. A., Aigbavboa, C. O., Thwala, D. W. and Edwards, D. J. 2019. Barriers to sustainable procurement in the Nigerian construction industry: an exploratory factor analysis. *International Journal of Construction Management*, 1-12.

[207] Gounden, K. 2016. *Factors influencing sustainable procurement within the private and public sector in South Africa*. University of Pretoria.

[208] Hasselbalch, J., Costa, N. and Blecken, A. 2015. Investigating the barriers to sustainable procurement in the United Nations. In: Klumpp, M., De Leeuw, S., Regattieri, A. and De Souza, R. (eds.) *Humanitarian Logistics and Sustainability.* Springer.

[209] Dutch Ministry of Infrastructure and the Environment .2017. *REBus Construction Lessons report.* Retrieved from https://bit.ly/2RIXUJX.

[210] ECOVADIS. 2021. *Sustainable procurement.* Retrieved from https://bit.ly/3xwFcYo.

[211] Australian Government. 1992. National strategy for ecologically sustainable development. In: *Department of The Environment and Energy* (ed.). Canberra, Australia. Retrieved from https://bit.ly/3GEh0qG.

[212] Australian Government. 2013. *Sustainable Procurement in the Australian Government Report 2013.*

[213] Environment Protection and Biodiversity Conservation Act 1999. No. 91, 1999. Retrieved from https://bit.ly/406REc3.

[214] ACT Government. 2015. *Sustainable procurement policy 2015*. Retrieved from https://bit.ly/3vx4GAT.

[215] SA state Government. 2012. *SA Sustainable Procurement Guideline*. Retrieved from https://bit.ly/3fqFGFX.

[216] Queensland Government. 2018. *Procurement OotCA (2018) Integrating sustainability into the procurement process*. Retrieved from https://bit.ly/3kG9PCQ.

[217] WA Department of Finance. 2017. *Guide to Sustainable Procurement 2017*.

[218] ACT Government (Directorate EaSD). 2011. *ACT waste management strategy towards a sustainable Canberra 2011–2025 reducing waste and recovering resources to achieve a sustainable, carbon-neutral* Canberra. Retrieved from https://bit.ly/3LHXBFN.

[219] NT EPA (Authority EP. 2015. *Waste management strategy for the Northern Territory* 2015–2022. Retrieved from https://bit.ly/3FadrXf.

[220] Department of Planning, Industry and Environment. 2021. *NSW waste and sustainable materials strategy 2041*. Retrieved from https://bit.ly/37fC9cs.

[221] Queensland Government. 2021. *Waste management and resource recovery strategy*. Retrieved from https://bit.ly/3OU8LJt.

[222] Tasmanian Government (Environment DoPIPWa). 2019. *Draft waste action plan* https://bit.ly/3wkkSr9.

[223] Waste Authority (Regulation DoWaE). 2020. *Waste Avoidance and Resource Recovery Strategy 2030: Western Australia's Waste Strategy*, https://bit.ly/3vHa5qj.

[224] Victoria State Government. 2020. Recycling Victoria- A new economy In: *Victoria Department of Environment, Land, Water and Planning* (ed.). Melbourne, Australia.

[225] eCOLOGIQ. 2020. Recycled First Policy. In: *Victoria Big Build*. Retrieved from https://bit.ly/3MFnU2J.

[226] King, S., Lusher, D., Hopkins, J. and Simpson, G. W. 2020. Industrial symbiosis in Australia: The social relations of making contact in a matchmaking marketplace for SMEs. *Journal of Cleaner Production*, 270, 122146.

[227] Waste Authority. 2020. Waste Avoidance and Resource Recovery Strategy 2030: Western Australia's Waste Strategy. In: *Regulation*, D. O. W. A. E. (ed.). Perth, Australia.

[228] Shooshtarian, S., Maqsood, T., Wong, P. S. and Bettini, L. 2022. Application of sustainable procurement policy to improve the circularity of construction and demolition waste resources in Australia. *Materials Circular Economy*, 4, 1-22.

[229] Merchant, L. (presenter). 2021. *Road to Reuse outcome*. Email communication.

[230] MainRoads WA (Cited in Shooshtarian, S., Le, T. J., Feng, Y. and Bettini, L. 2022. Analysis of sustainable procurement in supplying recycled content: A case study in Western Australia. *IOP Conference Series: Earth and Environmental Science*. 1101 (2022) 042004).

[231] Alex Fraser Group. 2016. *Dingley bypass*. Retrieved from https://bit.ly/3ybFlka.

[232] Victoria State Government. 2019. *Whole of Victorian government social procurement framework: Annual report 2018-19*.

[233] NSW Government. 2022. *Parramatta light rail*. Retrieved from https://bit.ly/37d6A2O.

[234] Cheng, W., Appolloni, A., D'Amato, A. and Zhu, Q. 2018. Green public procurement, missing concepts and future trends–A critical review. *Journal of Cleaner Production*, 176, 770-784.

[235] Wong, J. K. W., San Chan, J. K. and Wadu, M. J. 2016. Facilitating effective green procurement in construction projects: An empirical study of the enablers. *Journal of Cleaner Production*. 135, 859-871.

[236] Udawatta, N., Zuo, J., Chiveralls, K. and Zillante, G. 2015. Attitudinal and behavioural approaches to improving waste management on construction projects in Australia: benefits and limitations. *International Journal of Construction Management*, 15, 137-147.

[237] Sharma, H. B., Vanapalli, K. R., Cheela, V. S., Ranjan, V. P., Jaglan, A. K., Dubey, B., Goel, S. and Bhattacharya, J. 2020. Challenges, opportunities, and innovations for effective solid waste management during and post COVID-19 pandemic. *Resources, Conservation and Recycling*, 162, 105052.

[238] Australia Department of Health. 2020. *First confirmed case of novel coronavirus in Australia*. https://bit.ly/41lQxWQ.

[239] Khadem, N. 2020. Australians are hurting from the Coronavirus-led recession, but we fare better than most countries. *ABC News*. https://ab.co/3zNtrMW.

[240] Lv, H., Li, Y., Yan, H.-B., Wu, D., Shi, G. and Xu, Q. 2021. Examining construction waste management policies in mainland China for potential performance improvements. *Clean Technologies and Environmental Policy*, 23, 445-462.

[241] Van Fan, Y., Jiang, P., Hemzal, M. and Klemeš, J. J. 2021. An update of COVID-19 influences on waste management. *Science of the Total Environment*, 754, 142014.

[242] Mohammad, A., Goli, V. S. N. S. and Singh, D. N. 2021. Discussion on 'challenges, opportunities, and innovations for effective solid waste management during and post COVID-19 pandemic, by Sharma et al. (2020)'. *Resources, Conservation, and Recycling*, 164, 105175.

[243] Yeheyis, M., Hewage, K., Alam, M. S., Eskicioglu, C. and Sadiq, R. 2013. An overview of construction and demolition waste management in Canada: a lifecycle analysis approach to sustainability. *Clean Technologies and Environmental Policy*, 15, 81-91.

[244] Waste Advantage. 2020. The impact of COVID-19 on the waste and recycling industry. *Waste Advantage Magazine*. Retrieved from https://bit.ly/3mmG1A0.

[245] Klemeš, J. J., Fan, Y. V., Tan, R. R. and Jiang, P. 2020. Minimising the present and future plastic waste, energy and environmental footprints related to COVID-19. *Renewable and Sustainable Energy Reviews*, 127, 109883.

[246] Readling, A. 2020. C&D recycler survey shows businesses have been hit hard by COVID-19. *Waste Today US*. Retrieved from https://bit.ly/41k5OrP.

[247] Zambrano-Monserrate, M. A., Ruano, M. A. and Sanchez-Alcalde, L. 2020. Indirect effects of COVID-19 on the environment. *Science of the Total Environment*, 1(728):138813.

[248] US National Waste & Recycling Association. 2020. *NWRA, EREF survey highlights COVID-19's impact on the waste and recycling industry*. Retrieved from https://bit.ly/3uEXMvH.

[249] Love, BJ. and Rieland, J. 2020. COVID-19 is laying waste to many US recycling programs, *The Conversation*. Retrieved from https://bit.ly/3KGToBU.

[250] You, S., Sonne, C. and Ok, Y. S. 2020. COVID-19's unsustainable waste management. *Science*, 368(6498):1438-1438.

[251] Roberts, K. P. Stringfellow A. and Williams, I. 2020. Rubbish is piling up and recycling has stalled—waste systems must adapt, *The Conversation* https://bit.ly/3vd2Y93.

[252] ACOR. 2020. *Recycling and covid19 – some aspects to consider*. Retrieved from https://bit.ly/4147KEF.

[253] Zikmund, W. G., Carr, J. C. and Griffin, M. 2013. *Business Research Methods*, Cengage Learning, Belmont, CA, USA.

[254] Cooper D. R., Schindler P. S. and Sun J. 2006. *Business research methods*, New York, US. McGraw Hill eBook.

[255] Shooshtarian, S., Maqsood., T, Khalfan, M., Wong, S. P. and Yang, J. R. 2019. *Managing construction and demolition (C&D) waste in Australia' CIB World Building Congress 2019 'Constructing Smart Cities', Hong Kong, China*. 17-21 June 2019.

[256] National Health and Medical Research Council. 2007. *Australian National Statement on Ethical Conduct in Human Research* https://bit.ly/3rpVZIQ.

[257] Di Gregorio, S. 2000. *Using Nvivo for your literature review*. Retrieved from https://bit.ly/3o9Srvu.

[258] Braun, V. and Clarke, V. 2006. Using thematic analysis in psychology. *Qualitative Research in Psychology*. 3(2):77-10.

[259] ABS. 2020. *Gender Indicators*, Australia. Retrieved from https://bit.ly/409Fm2x.

[260] A payment to help keep Australians in jobs and supported businesses affected by the significant economic impact of the COVID-19 pandemic.

[261] *Waste Management Review*. 2020. Industry responds to COVID-19 support packages. Retrieved from https://bit.ly/414UijS.

[262] The NSW EPA. 2020. *Information fact sheet*.

[263] ABS. 2021. *National, state and territory population*. Retrieved from https://bit.ly/3hwzWM0.

Index

T

W

About the Authors

Dr. Salman Shooshtarian holds a PhD in the Built Environment and is currently working as a Lecturer at RMIT University. He is an active researcher in the fields of sustainability, circular economy, and C&D waste management. He has experience in designing and performing empirical research projects and has established collaboration with different research teams for several research projects. He is the lead author of a book titled construction and demolition waste management in Australia (2021).

Professor Tayyab Maqsood has been involved in research related to C&D waste management for over a decade. He is leading the C&D waste management research group at RMIT and is the founder of Australia's first National Construction and Demolition Waste Research and Industry Portal. He is a strong proponent of the circular economy or cradle-to-cradle approach. He is also Associate Dean of Project Management in the School of Property, Construction and Project Management at RMIT University.

Dr. Savindi Caldera is a Lecturer at University of the Sunshine Coast and an Adjunct Research Fellow at the Cities, Research Institute, Griffith University. As an expert in lean and green manufacturing, cleaner production and rapid capacity building, she has been working with public and private sector colleagues to enable resilient natural and built environments. She has published over 50 publications and has secured research funding over AUD$2.1 million.

Professor Tim Ryley is Professor of Aviation, within the School of Engineering & Built Environment, Griffith University. His research expertise principally covers air transportation and environmental issues, but he has led a series of multi-disciplinary transport and engineering-based projects, including C&D waste management and demand responsive transport. Prof. Ryley has led or co-led research projects totalling over $5 million, typically funded by research councils and industry.

Professor Peter S. P. Wong is the Associate Dean of the School of PCPM. As of 2022, he has more than 100 publications and has been awarded over AUD $1.8 million research grants. Professor Wong is a chartered construction manager and a chartered quantity surveyor with solid industry experience. His knowledge and experience particularly in organisational and operational management in construction provides a fundamental basis for effective contribution to research projects in relevant fields.